Recalcitrant Seeds

Recalcitrant Seeds:
Causes and Effects

Authors
K. Sivasubramaniam, Ph.D.
K. Raja, Ph.D.
R. Geetha, Ph.D.

2012

SSPH

SATISH SERIAL PUBLISHING HOUSE
403, Express Tower, Commercial Complex, Azadpur, Delhi-110033 (India)
Phone : 011-27672852, Fax : 91-11-27672046
E-mail :info@satishserial.com, hkjain1975@yahoo.com
Website : www.satishserial.com

Published by :

SATISH SERIAL PUBLISHING HOUSE
403, Express Tower, Commercial Complex, Azadpur, Delhi-110033 (INDIA)
Phone : 011-27672852 Fax : 91-11-27672046
E-mail : info@satishserial.com, hkjain1975@yahoo.com

ISBN 978-93-81226-00-1

Typeset at: Harminder *for* Laxmi Art Creation
Printed at: Salasar Imaging Systems, Delhi-110035

Preface

Recalcitrant (non-orthodox, homoeohydrous) seeds are a phenomenon into themselves. Several of the species from mesic forest and few from temperate highlands and wetlands show recalcitrant behaviour. It is a fact, that several of the horticultural fruits of economic importance are recalcitrant as well as a few oil bearing trees. Though orthodox behaviour promoted agrarian settlements, recalcitrant seeds were also used for their pulp bearing fruits, hence domesticated side by side along with agricultural crops. Research into recalcitrance had shown a behaviour ranging from complete orthodox, to some semblance of recalcitrant behaviour, to complete recalcitrancy, prompting the high priestess of recalcitrant phenomenon, Dr. Patricia Berjak to opine that it is continuum. Even among the same genera, species show varied degree of recalcitrant behaviour depending on the habitat they occupy, meaning, recalcitrant behaviour is a method of adaptation. Still in its infancy, research into recalcitrant behaviour is also a continuum. With several of plant species being lost to human related activities and several more as a consequence of global warming and climatic changes, it is high time, plant species especially in tropical rainforest were categorized for their recalcitrance behaviour and protocols evolved for their storage. This book is a result of intense scrutiny into recalcitrant literature, thus may be exhaustive and useful to fellow scientists and teachers. The authors however would caution the reader, about the necessity to use this book as the first step into the world of recalcitrant phenomenon. It may serve as a beacon into future research and help them to appreciate the world of nature.

K.Sivasubramaniam
K.Raja
R.Geetha

Contents

Chapter 1

Introduction

Generally the seeds are classified into "orthodox" and "recalcitrant" based on their desiccation and temperature sensitivity. Most of the agricultural and horticultural crop seeds come under orthodox class. These seeds have the ability to tolerate reduction in moisture content and temperature, which increase the life span of seeds. It is opined that the orthodox storage nature of most agricultural crops helped man to cultivate land (Barbedo and Bilia, 1998) and should have been co-responsible for evolution of agriculture. Orthodox seed behaviour would allow seeds to be stored and sown at different times of the year, thus helping humans to cultivate crops when ever there was need and not depend on nature to provide rain to raise crops. This orthodox nature was further refined by humans, such that they could withstand desiccation. Further, these seeds facilitated storage thus favoring human settlement. Wars and alliances were carried out to raid stored food and forge alliances in sharing the stored seeds.

These groups of seeds remain viable for longer periods even upto hundreds of years when they are dried and stored properly. They can be even stored at very low temperature of –196°C. Recalcitrant (non-orthodox) seeds on the other hand, are killed if their moisture contents are reduced below some relatively high critical value of 20 to 35%. It is estimated that 15% of the world's flora posses recalcitrant seeds i.e., approximately 37,500 species. In horticultural crops, except vegetables, few spices and fruits, most of the seeds show the recalcitrant behaviour. Several species of the *Dipterocarpaceae* family that yield valuable timber belong to recalcitrant species underlying their economic importance in conservation and preservation of this fragile endowment provided by nature (Chin, 1978).

Generally, recalcitrant seeds occur in humid forest environment. They may also occur in tropical, subtropical or temperate regions. The fruit crops viz., mango (*Mangifera indica*), few species in citrus (*Citrus* spp.), nutmeg (*Myristrica fragrans*), avocado (*Persea americana*), jack (*Artocarpus heterophyllus*), jamun (*Syzygium cuminii*), litchi (*Litchi chinensis*), loquat (*Eribotrya japonica*), mangosteen (*Garcinia mangostana*), durian (*Durio zibethinus*), longan (*Euphoria longan*), rumbutan (*Nephelium lappaceum*) and curry leaf (*Murraya koenigii* (Linn.) Sprengel) come under the recalcitrant seed category (Doijode, 2001). The plantation crops like (*Areca catechu*), coconut (*Cocos nucifera*), cocoa (*Theobrama cacao*), coffee (*Coffea* spp.), clove (*Eugenia aromaticum*), nutmeg (*Myristrica fragrans*), rubber (*Hevea brasiliensis*) and tea (*Camellia sinensis*) and spices like pepper (*Piper nigrum*), curryleaf, cinnamon (*Cinnamomum verum*) and cassia (*Cinnamomum aromaticum*) are also some of the examples of recalcitrant seeds. Forest tree species belonging to the *Araucariaceae* and *Dipterocarpaceae* posses recalcitrant behaviour (Chin, 1989). An exhaustive list of available recalcitrant as well as viviparous seeds was listed out be Farnsworth (2000) along with their habitat (Table 1.1.)

Table 1.1 : List of plants showing recalcitrant / viviparous behaviour (Farnsworth, 2000)

Family	Species	Life form	Latitude	Habitat	Seed status
Aceraceae	*Acer spp.*	T	T	Mesic forest	R
Alismataceae	*Sagataria latifolia*	H	T	Wetland	R
Anacardaceae	*Mangifera indica*	T	TR	Wet forest	R
Annonaceae	*Cymbopetalum baillonii*	ST	TR	Wet forest	R
Apocynaceae	*Hancornia speciosa*	T	TR	Wet forest	R
	Landophia kirkii	V	TR	Wet forest	R

Contd...

Araceae	*Dieffenbachia longispatha*	S	TR	Wet forest	V
Araceae	*Xanthosoma sagittifolim*	S	TR	Swamp forest	R
Araliaceae	*Hedera helix*	V	T	Mesic cultivar	R
Araucariaceae	*Agathis robusta*	T	TR	Wet forest	R
	Araucaria spp.	T	TR	Coast	R
Arecaceae	*Areca catheca*	P	TR	Wet forest	R
	Calmus spp.	P	TR	Swamp forest	R
	Chrysalidocarpus leutecens	P	TR	Wet forest	R
	Cocos nucifera	P	TR	Coast	V
	Elaeis guineensis	P	TR	Wetforest (ag)	R
	Nypa fruticans	P	TR	Coast	V
	Sabal spp.	P	TR	Wet forest	R
Asteraceae	*Abrotanella linearis*	S	ST	Wet forest	V
	Acamptopappus sp.	S	ST	Desert	R
Avicenniaceae	*Avicennia spp.*	T	TR	Coast	V
Bombacaceae	*Durio zibethinus*	T	TR	Wet forest	R
	Montezuma speciosissima	T	TR	Wet forest	V
Boraginaceae	*Cordia allidora*	T	TR	Wet forest	R
Buseraceae	*Daryodes excelsa*	T	TR	Wet forest	R
Caricaeae	*Jacaratia doichaula*	ST	TR	Wet forest	R
Caryophyllaceae	*Scheidea diffusa*	W	TR	Montane	V
Celastraceae	*Salaciopsis ingifera*	V	TR	Wet forest	V
Ceratophyllaceae	*Ceratophyllum sp.*	H	TR	Revirine	R
Chenopodiaceae	*Chenopodium quinoa*	H	T	Agriculture	R
Chrysobalan-aceae	*Coupeir polyandra*	ST	TR	Wet forest	R
Clusiaceae	*Garcinia mangostana*	ST	TR	Wetforest (ag)	R
	Symphonia globulifera	ST	TR	Wet forest	R
Combrestaceae	*Conocarpus erectus*	ST	TR	Coast	R
	Laguncularia racemosa	T	TR	Coast	R
Connaraceae	*Connarus grandis*	ST	TR	Wet forest	V

Contd...

Cornaceae	*Corokia macrocarpa*	S	ST	Wet forest	V
	Griselinia spp.	S	ST	Wet forest	V
Corylaceae	*Corylus Americana*	ST	T	Mesic forest	R
Cucurbitaceae	*Sechium edule*	H	T	Agriculture	V
	Telfaira occidentalis	S	TR	Agriculture	R
Cupressaceae	*Cupressus macrocarpa*	ST	TR	Wet forest	R
Cymodoceaceae	*Amphibolus spp.*	H	TR	Coast	V
Cymodoceaceae	*Thalassodendron spp.*	H	TR	Coast	V
Dipterocarpaceae	*Anisoptera laevis*	T	TR	Wet forest	R
	Dipterocarpus spp.	T	TR	Wet forest	R
	Dryobalanos aromatica	T	TR	Wet forest	R
	Hopea spp.	T	TR	Wet forest	R
	Parashorea densiflora	T	TR	Wet forest	R
	Shorea robusta	T	TR	Wet forest	R
	Stemonoporus oblongifolius	T	TR	Wet forest	R
Ebenaceae	*Diospyros virginiana*	ST	ST	Mesic forest	R
Elaeocarpaceae	*Sloanea berteriana*	T	TR	Wet forest	R
Euphobiaceae	*Heavea brasiliensis*	T	TR	Wet forest	R
Fabaceae	*Castanospermum australe*	T	TR	Coast	R
	Inga spp.	ST	TR	Swamp forest	V
	Mora oleifera	T	TR	Coast	V
	Pithecellobium racemosum	T	TR	Wet forest	V
Fagaceae	*Castanea dentata*	T	T	Mesic forest	R
	Lithocarpus densiflorus	T	T	Mesic forest	R
	Quercus spp.	T	T	Mesic forest	R
Flacourtiaceae	*Casearia corymbosa*	T	ST	Wet forest	R
	Dovyalis hebecarpa	ST	T	Agriculture	R
	Flacourtia indica	ST	TR	Wet forest	R
	Muntingia calabura	ST	TR	Wet forest	R
Hippocastanaceae	*Aesculus hippocastanum*	ST	T	Mesic forest	R
Lauraceae	*Machilus thunbergii*	ST	TR	Swamp forest	R
	Nectandra ambigens	ST	TR	Wet forest	R
	Persea americana	S	ST	Wet forest	R

Contd...

Lecythidaceae	*Barringtonia racemosa*	T	TR	Coast	V
	Bertholletia excelsa	T	TR	Wet forest	R
	Lecythis ampla	T	TR	Wet forest	R
Liliaceae	*Crinum capense*	H	TR	Riverine	V
	Hymenocallis spp.	H	ST	Riverine	V
	Nerine sp.	H	TR	grassland	V
	Ripogonum scandens	H	ST	Wet forest	V
Lobeliaceae	*Lobelia sp.*	S	TR	Montane	R
Loganiaceae	*Fagraea fragrans*	S	TR	Wet forest	R
Magnoliaceae	*Magnolia portoricensis*	ST	TR	Wet forest	R
	Michelia champaca	T	TR	Wet forest	R
Melastomataceae	*Melastoma malabathricum*	S	TR	Swamp forest	R
Meliaceae	*Aglaia odorata*	S	TR	Wet forest	R
	Carapa guianensis	T	TR	Swamp forest	R
	Guarea glabra	T	TR	Wet forest	R
	Turrianthus africana	T	TR	Wet forest	R
Moraceae	*Artocarpus heterophyllus*	T	TR	Wet forest	R
	Morus latifolia	ST	ST	Wet forest	V
Myristicaceae	*Myristica hollrungii*	T	TR	Swamp forest	V
	Aegiceras spp	ST	TR	Coast	V
Myrtaceae	*Amomyrtus lama*	T	ST	Wet forest	R
	Eugenia spp.	ST	TR	Wet forest	R
Nepenthaceae	*Nepenthes gracilis*	E	TR	Swamp forest	R
Nyctaginaceae	*Pisonia longirostris*	T	TR	Swamp forest	V
Nymphaceae	*Nymphaea sp.*	H	TR	Riverine	R
Nyssaceae	*Nyssa aquatica*	T	ST	Swamp forest	R
Oxalidaceae	*Averrhoa carambola*	ST	TR		R
	Oxalis sp.	S	T	Mesic forest	R
Pellicierieaceae	*Pelliciera rhizophorae*	T	TR	Coast	V
Piperaceae	*Piper hispidum*	S	TR	Wet forest	R
Plumbaginaceae	*Aegialitis spp.*	S	TR	Coast	V
Poaceae	*Spartina anglica*	H	T	Coast	R
	Zizania aquatica	H	T	Wetland	R

Contd...

Podocarpaceae	*Dacrycornus dacrydioides*	ST	ST	Mesic forest	R
Podocarpaceae	*Podocarpus henkelii*	S	ST	Mesic forest	R
Polygonaceae	*Fagopyrum esculentum*	H	T	Agriculture	V
Potamogentonaceae	*Potamogeton sp.*	H	T	Wetland	R
Proteaceae	*Macadamia ternifolia*	ST	TR	Agriculture	R
Ranunculaceae	*Caltha palustris*	H	T	Wetland	R
Rhizophoraceae	*Bruguiera spp.*	T	TR	Coast	V
	Carallia brachiata	ST	TR	Swamp forest	R
	Ceriops spp.	ST	TR	Coast	V
	Kandelia candel	ST	TR	Coast	V
	Rhizophora spp.	T	TR	Coast	V
Rosaceae	*Eriobotrya japonica*	ST	T	Mesic forest	R
Rubiaceae	*Coffea spp.*	ST	TR	Montane	R
	Coprosma robusta	S	ST	Wet forest	V
	Ixora sp.	ST	TR	Wet forest	R
	Ophiorrhiza tomentosa	S	TR	Wet forest	V
	Posoqueria latifolia	ST	TR	Swamp forest	R
Rutaceae	*Citrus spp.*	ST	TR	Agriculture	R
	Clausena dentata	ST	TR	Agriculture	R
	Fortunella japonica	ST	TR	Agriculture	R
Santalaceae	*Santalum album*	ST	TR	Wet forest	R
Sapindaceae	*Euphoria longan*	ST	TR	Wet forest	R
	Litchi chinensis	ST	TR	Wet forest	R
	Magonia pubescens	ST	TR	Wet forest	R
	Meliococcus bijugatus	ST	TR	Wet forest	R
	Nephelium lappaceum	T	TR	Wet forest	R
Sapotaceae	*Calocarpum sapota*	T	TR	Wet forest	R
	Chrysophyllum cainito	ST	TR	Wet forest	R
	Manilkara zapota	ST	TR	Wet forest	R
	Mimusops sp.	T	TR	Agriculture	R
	Pouteria ramiflora	T	TR	Wet forest	R
	Euphrasia disperma	H	ST	Wet forest	V

Contd...

Scrophulariaceae	*Quassia indica*	T	TR	Wet forest	R
Simaroubaceae	*Cola nitida*	ST	TR	Mesic forest	R
Sterculiaceae	*Theobroma cacao*	ST	TR	Wet forest	R
Surianaceae	*Guilfolylia monostylis*	T	TR	Wet forest	R
Theaceae	*Camellia sinensis*	ST	TR	Montane	R
Verbenaceae	*Vitex divaricata*	T	TR	Wet forest	R
Vochysiaceae	*Vochysia honurensis*	T	TR	Wet forest	R

Life form of the adult plant (T : tree taller than 10 m; ST : small tree; S : shrub; V : vine / liana; P : palm; H : herbaceous);

Latitude or native region of the species (T : temperate; TR : tropical; ST : sub-tropical);

Habitat of the species, including agricultural ("ag") of the species is cultivated;

Seed status (R : recalcitrant; V : viviparous).

1.1. Seed Recalcitrance

Recalcitrance is the phenomenon of viability loss shown by some seeds to moisture loss. These seeds are well known for their sensitivity to desiccation and freezing temperatures. Harrington (1972) proposed a basic rule related to the effect of seed moisture content and storage temperature in which the reduction in moisture content and temperature increases the life span of seeds. The famous Harrington's thumb rule quotes as "for every 1% reduction in seed moisture content and for every 5°C reduction in storage temperature, the seed life is doubled". This rule holds good for moisture content between 5-14% and 0-50°C. Obeying the above rule, most seeds remain viable for longer periods when they are dried and stored at low temperatures, even as low as –196°C. But, Roberts (1973) clarified the situation by introducing the terms 'orthodox' and 'recalcitrant' to describe the aberrant storage behaviour of some seeds. He referred to those seeds obeying Harrington's rule as seeds that can tolerate desiccation and freezing temperatures namely 'orthodox', while the other types of seed that are readily killed by desiccation when the moisture content falls below a critical value between 12 to 31% and cannot tolerate freezing temperatures as 'recalcitrant'. Later, Hanson (1984) suggested more accurate and descriptive terms, by which orthodox seeds are called 'desiccation-tolerant' and recalcitrant seeds are referred to as 'desiccation-sensitive'.

Pammenter *et al.* (1991) postulated the difference between desiccation tolerant and intolerant seeds was their differences in response to loss of non-freezable water rather than the difference between seeds in the amount of non freezable water *per se.* While Kermode (1990) opined that orthodox seeds undergo desiccation upon maturation, an essential step that changes their metabolism from development to germination mode. In recalcitrant seeds such 'switching off' metabolism is not noticed due to factors like hormonal balance, presence of sugars and proteins, physical properties of seed borne water.

1.2. Basis for Recalcitrance

Early research attempted in Brazil by Castro and Krug (1951) with *Inga edulis* and *Inga striata* seeds reported that seeds did not rest after maturity and started to develop and germinate inside the fruit itself and could not withstand drying. This enabled them to provide the two essential characteristic of recalcitrance – 'vivipary' and 'desiccation intolerance'. Bacchi (1961) recorded that seeds of *Inga edulis* lost viability when moisture content was reduced beyond 35% and expounded the theory that such seeds were short lived. Hence, recalcitrant seeds are those that undergo little or no maturation drying and remain desiccation sensitive both during development and after shedding.

Barbedo and Bilia (1998) concluded that recalcitrant seeds were selected in a habitat distinct from orthodox seeds, which permitted seedling development all through the year. Further, the moist / humid environment prevented maturation drying. Under moist humid rainforests, seeds if dried could not germinate and would loose-out to competition from those that had already germinated. Hence, evolution saw to that only those species that were ready to germinate upon embryo development survived, while those that dried eventually died.

Recalcitrant seeds thus, are shed hydrated but the water content will be between 0.43-4.0 g g^{-1} that is 30-80% on wet weight basis. The shedding moisture content of seeds is characteristic of the species involved and is correlated with degree of desiccation developed by individual species (Finch-Savage, 1996). The presence of sufficient water to stimulate germination upon shedding is an essential feature for continued embryo development (increasing dry weight) in all recalcitrants. This occurs in *Podocarpus henkelii* seeds (Dodd and Van

Staden, 1981), some dipterocarps (Tompsett, 1987) and *Cycas revoluta* (Dehgan and Schutzman, 1989).

Berjack and Pammeter (1994) opined that there were two different factors that operate in orthodox and recalcitrant seeds viz., 1. damage that occurs at low water contents when non-freezable water (water that is held to stabilize intercellular structures and macromolecules) is removed and is termed as 'desiccation damage' and 2. damage that occurs during slow dehydration, during which, metabolic imbalances are reasoned to cause the generation of free radicals, termed as 'metabolic damage'. Hence desiccation damage is attributed to the lack or inadequate operation of the processes / mechanisms that provide protection to desiccation tolerant seeds in the dry state, while metabolic damage occurs in truly recalcitrant seeds that are denied the protection by slew of mechanisms / processes that facilitate the acquisition and maintenance of desiccation-tolerance as exhibited by maturing and mature orthodox seed types.

The declining fluidity of the membrane lipids during chilling is also one of the reasons for recalcitrance (Wolfe, 1978). King and Roberts (1979) suggested the following two hypothesis by which death occurs in these seeds i.e., 1. Either death occurs rapidly at / or below some critical moisture content (critical moisture content hypothesis), or 2. Loss of viability occurs at a rate which is negatively related to moisture content over a wide range of moisture contents (non critical moisture content hypothesis). Similarly, Pammenter and Berjak (1999) opined that the damage of recalcitrant seeds were due to absence or incomplete expression of physical characteristics of cells and intracellular constituents, metabolic 'switching off' and 'switching on' mechanism, inefficient operation of antioxidant systems, accumulation of putatively protective substances including the Late Embryogenic Abundant (LEA) proteins, the presence and operation of repair systems, sucrose and other oligosaccharides as well as amphipathic molecules and the presence of oleosins. Earlier, scientists have suggested the possible reasons for fall in viability of recalcitrant cocoa seeds which include (i) the presence of some temperature dependent, rate limiting reaction, the cessation of which causes lethal metabolic disruption; (ii) the absence of some protective substance which is present in those seeds not susceptible to chilling; and (iii) the liberation of some toxic material owing to cold induced changes in membrane permeability (Boroughs and Hunter, 1963).

All orthodox seeds are able to tolerate dehydration to 5% (0.053g H_2O g^{-1}) of seed dry weight even when maturation drying is not completed prior to shedding. Further, orthodox seeds withstand loss of water sorbed to macromolecular structures, particularly the membrane surfaces, without irreversible denaturation (Clegg, 1986 and Crowe and Crowe, 1986). In orthodox seeds, the water associated with membranes is replaced by polyhydroxyl compounds that stabilize the macromolecules as dehydration takes place. Such replacement does not take place in recalcitrant seeds. Berjack and Pammmeter (1994) thus referred to those seeds that do not tolerate afore mentioned condition as 'non-orthodox'. Such seeds have been described as recalcitrant (Roberts, 1973) or intermediate (Ellis *et al.*, 1990). Later Pammenter *et al.* (1991) coined the term 'Homeohydrous' seeds to indicate the constant presence of seed moisture.

Farrant *et al.* (1988) opined that the desiccation sensitivity of recalcitrant seeds to water loss leading to irrecoverable damage varies among species and there are degrees of desiccation sensitivity with some species that cannot with stand even slight desiccation while others can be tolerant to desiccation, thus exhibiting a sort of 'continuum' of recalcitrant behavior from species that are highly desiccation sensitive to that of those that will tolerate drying to the lowest water content but behave like recalcitrant seeds when exposed to low temperatures.

Farrant *et al.* (1988) further characterized recalcitrant seed behaviour as minimum, moderate and high based on their ability to tolerate desiccation and temperature sensitivity (Table 1.2.).

Table 1.2 : Types of recalcitrance

Parameter	Minimum	Moderate	High
Desiccation sensitivity	Fair amount of water loss tolerated	Moderate amount of water loss tolerated	Little water loss tolerated
Germination rate	Slow in absence of additional water	Moderate in absence of additional water	Fast in absence of additional water
Temperature sensitivity	Lower temperature tolerated	Most species are temperature sensitive	Most species are temperarture sensitive.
Habitat	Temperate / tropical	Tropical	Tropical forests and wetland
Examples	Quercus, Araucaria	Theobroma, Hevea	Avicennia, Rhizophora

When assessing recalcitrancy, factors like hydrated storage lifespan and chilling sensitivity must be considered. The continuum grades from extreme desiccation sensitive types through the minimally recalcitrant types, to the intermediate seed species (a.k.a. difficult recalcitrant) to those that do not react adversely to low temperatures, through to those that are chilling sensitive when dehydrated (Hong and Ellis, 1996) and finally to orthodox seeds that will tolerate less or more extreme dehydration (Vertucci and Roos, 1990). Such factors made Berjak and Pammenter (1994) to comment that recalcitrance is not an 'all-or-nothing' situation and other factors than mere assessment of response to desiccation or chilling sensitivity must be considered before a seed is labeled as recalcitrant. Berjak and Pammenter (2008) later implied that the phenomenon was a open ended continuum of seed behaviour with extreme orthodoxy at one end and highest degree of recalcitrance at the other as species that are orthodox in one region are recalcitrant in another indicated the strong influence of habitat.

Evolutionarily speaking, the phenomenon of recalcitrance may be linked to bitegmic, crassinucellate ovules that have nuclear endosperm development noticed in 45 families of dicots. Pachychalazy (large seededness) and recalcitrance are considered as ancestral traits of dicots. The integuments and chalaza of recalcitrant ovules are highly vascularized, the chalaza is massive and surrounds the embryosac to varying degrees (von Teichman and van Wyk, 1994). This pachychalazy condition and vasularization facilitates moment of nutrient to the embryo till the end of seed dehiscence. Thus the absence of post vasucular separation phase seen in orthodox seeds is abscent in recalcitrant seeds. Hence, due to vascularization and accumulation of nutrients, the seeds of advance dicots are small and possess little or no vascularization. Further, most recalcitrant seeds are exalbuminous, thus transferring the storage function to embryo. Transfer of nutrient to cotyledons or embryo (consequently seen as large) in recalcitrant species, facilitates quicker germination. Starch or soluble sugar constitutes the main stored food in recalcitrant species while it is dehydrins and LEA proteins in orthodox seeds. While intermediate seeds store lipid (*Coffea arabica, Carica papaya, Elaeis guineensis, Azadirachta indica*) which could be the reason for varied degrees of recalcitrant behaviour of these intermediate types (Farrant *et al.*, 1988 and Berjak *et al.*, 1989).

Hence, seeds can thus be classified as orthodox, intermediate and recalcitrant. The cause for such behavior can be traced to the physiological status of seeds at various water potentials (Vertucci and Roos, 1990; Vertucci, 1993 and Vertucci and Farrant, 1995) and the properties of water at various hydration levels corresponding to specified water potential ranges (Vertucci and Farrant, 1995 and Vertucci, 1993).

Berjack and Pammenter (1994) further clarified and summarized that the degree of recalcitrant behavior could be attributed to the intracellular physical characteristics such as :

- reduction of the degree of vacuolation,
- amount and nature of insoluble reserves accumulated,
- integrity of the cytoskeleton,
- conformation of the DNA, chromatin and nuclear architecture.

While that of orthodox seeds as :

- Intracellular de-differentiation, which effectively results in the minimization of surface areas of membranes and probably also of the cytoskeleton,
- 'Switching off' of metabolism prior to harvest,
- Presence and efficient operation of antioxidant systems,
- Accumulation and roles of putatively protective molecules, including late embryogenic accumulating / abundant proteins (LEA's), sucrose and certain oligosaccharides or galactosyl cyclitols,
- Deployment of certain amphipathic molecules,
- An effective peripheral oleosin layer around lipid bodies,
- The presence and operation of repair mechanisms during rehydration.

❑❑❑

Chapter 2

Properties of Recalcitrant Seeds

Recalcitrant seeds are generally bigger than most agriculturally useful crops, shed with high moisture and difficult to dry and store. The characteristics of recalcitrant seeds are summarized in this chapter.

2.1. Physical Properties

Generally recalcitrant seeds are large in size and weigh more as compared to orthodox seeds because of their high moisture content. Many recalcitrant seeds are covered with a fleshy layer, which is often edible, as in case of avocado, durian, jackfruit, jamun, litchi, mango, mangosteen and rambutan. Similarly some recalcitrant seeds are large and are found in a single-seeded, simple drupes either wet eg. mango, durian, avocado or dry eg. arecanut and coconut or as seeds in a composite fruit such as jackfruit. The shape and size of recalcitrant seeds of different species differ vastly even within species or within the same fruit (Chin, 1989).

The main characteristics of orthodox and recalcitrant seeds are summarized in Table 2.1.

Table 2.1 : Comparison of orthodox and recalcitrant seeds

Characteristics	Orthodox	Recalcitrant
1. Tolerance to drying	Can be dried to low moisture content	Cannot be dried beyond critical low moisture level
2. Tolerance to desiccation	Tolerant	Sensitive
3. Tolerance to low temperature	Tolerant	Sensitive
4. Size of the seeds	Small to medium	Large
5. Storage life	Many years	Few days or few months
6. Storage methods	Ordinary storage	Special methods required
7. Seed moisture content at harvest	Low	High
8. Examples	Cereals, Millets, Pulses, Oil seeds, Vegetables.	Mango, Jack, Jamun, Arecanut, Coconut, Rubber

High initial (shedding) moisture content is also one the characteristic feature of recalcitrant seeds as provided in the Table 2.2. for some species.

Table 2.2 : Initial moisture content of some recalcitrant species.

Species	Initial moisture content (%)
Hopea odorata	52.0
Azadiracta indica	38.1
Pometia pinnata	42.6
Vitellaria paradoxum	40.7
Artocarpus heterophyllus	55.89
Quercus leucotrichophora	38.17
Mangifera indica (c.v.Alphonso)	64.0
Mangifera indica (c.v.Totapuri)	53.0
Syzygium cuminii	48.5

Chin *et al.* (1989) compiled the physical characteristics of some recalcitrant seed in comparison to orthodox seed. It is seen that seed size of recalcitrant seed was larger and the thousand seed weight was also high. Similarly the moisture content was very high in comparison to orthodox seeds (Table 2.3.).

Table 2.3 : Comparison of recalcitrant and orthodox seed physical characteristics

Crop	Seed size (LxB) (mm)	1000 seed weight (g)	Moisture content g water /g fresh wt
Recalcitrant			
Nephelium lappaceum L	28 x 16	3555	0.49
Artocarpus heterophyllus Lam	35 x 24	8520	0.52
Artocapus champeden Spreng.	30 x 20	5814	0.71
Lansium domesticum Corr	17 x 13	2335	0.52
Bouea ganadaria	22 x 15	3530	0.46
Durio zibethinus Murr.	42 x 25	14783	0.50
Theobroma cacao L	25 x 25	1995	0.36
Orthodox			
Hibiscus esculentus L	6 x 4	146	0.18
Vigna sesquipedalis (L) Fruw	12 x 5	192	0.16

Chandel *et al.* (1995) reported the physical characteristics of recalcitrant seeds as large size, higher fresh weight, high seed moisture, large embryonic axis and high moisture content in embryonic axis (Table 2.4.) in comparison to orthodox seed.

Table 2.4 : Seed and embryo characteristics of some recalcitrant species in comparison to orthodox species (French bean) Chandel *et al.* (1995)

Character	Tea	Cocoa	Jackfruit	French bean
Seed size (L x B) (cm)	0.9 x 1.2	3.0 x 1.5	2.5 x 1.9	1.4 x 0.7
Fresh weight of 100 seeds (g)	187.0	124.2	359.8	41.00
Seed moisture Content (% fwb)	40-43	40-45	50-56	11-13
Size of embryonic axis (L x B) (cm)	0.21 x 0.12	0.8 x 0.3	0.4 x 0.3	0.55 x 0.1
Axes moisture Content (% fwb)	50-52	65-70	65-68	12-13

It is easily evident that except of tea, seeds of crops like cocoa and jackfruit are twice the length and width of an orthodox seed. The fresh weight was almost 4-8 times the size of normal orthodox seed with moisture content being 4-5 times of French bean. Similarly the embryonic axes were relatively larger but possessed 4-5 times the moisture. Thus recalcitrant seeds were large and had high moisture content with large embryonic axis, which was metabolically active and possessed high moisture thus making it more vulnerable when desiccated.

Several of the wild, as well as cultivated recalcitrant species also exihibit poly emryony. Jamun shows polyembryony to an extent of 20-50% as well as in mangosteen, where seeds are formed from nucellar tissues of carpel. This compounded the problem in assessing germination and generating seedlings from true embryo.

2.2. High Shedding Moisture Content and Critical Moisture Content

As discussed earlier, recalcitrant seeds are well known for their sensitivity to desiccation, especially large seeded species generally found in the tropics. In recalcitrant seeds, decline in viability occurs abruptly below a certain moisture level, which is called "Critical Moisture Content (CMC)" (Poulsen and Eriksen, 1992) at which 50% of the seeds lost viability (V_{50}). Tompsett (1986) coined the term "Lowest Safe Moisture Content (LSMC)" which defined as the moisture content below which freshly collected seeds died when the seed lot is dried.

The other important property of recalcitrant seeds is their high moisture content, even after they have been shed from the mother plant. Unlike orthodox seeds, they do not undergo maturation drying. These recalcitrant seeds generally have high moisture content, ranging from 30 to 70% (Chin, 1989), eg., the freshly collected arecanut drupe and seed has 63.6% and 50.1% moisture content respectively. However, unlike orthodox seeds arecanut seeds are highly intolerant of further desiccation and thus classifying them as recalcitrant (Das and Ray, 1985; Nagwekar *et al.*, 1997). Delgado and Barbedo (2007) researching with six species of genus *Eugenia* (E. *brasiliensis* Lam., E. *cerasiflora* Mig., E. *involucrate* D.C., E. *pyriformis* Camb., E. *umbeliflora* Berg., and E. *uniflora* L.) found that all the seeds started loosing viability when moisture level was dropped below 45% and completely below 15%. Lack of endosperm to protect the seed from moisture loss was attributed as the cause of poor storability in *Ulmus laevigata* Royle seed (Bhardwaj *et al.,* 2001b).

Moisture is a decisive factor in maintenance of recalcitrant seed quality during storage. In these seeds, the sub-cellular water is strongly associated with macromolecular surfaces, assuring the stability of membranes and macromolecules. Water loss during the drying process causes a series of metabolic alterations, acting on the control of growth regulators, amount and types of proteins and sugars, presence of free radicals, physical water status, among others, resulting in the on-set of the deterioration process. This moisture was termed as critical moisture content when a significant reduction in germination percent was observed and as lethal moisture content when no germination occurs (Cibele *et al.,* 2003).

Again, Tompsett (1987) stated that the critical moisture content as the lowest safe moisture content. The degree of sensitivity varies between species (Roberts, 1973) but generally the critical moisture content is 20-35% below which the seeds will be killed. The critical moisture content for "mother of clove" (*Syzygium aromaticum* L) or clove seed was 35%, 30% for nutmeg (*Myristica fragrans* Houtt.) and 17-25% for rubber (*Havea brasiliensis* Muell-Arg.) (Gunasekaran, 1997). A list of critical moisture contents for some species is provided below (Table 2.5.). It can be observed, that the critical moisture content varies among varieties within the same species.

Table 2.5 : Critical moisture content of few recalcitrant seeds

	Crop seeds	Critical Moisture Content (%)	References
1.	*Quercus robur* (Oak)	38	Vlase, 1970
2.	*Nephelium lappaceum*	20	Chin, 1975
3.	*Shorea talura*	17	Sasaki, 1976
4.	*Hopea helferi*	35	Tamari, 1976
5.	*Hevea brasiliensis* (Rubber)	15-20	Chin *et al.*, 1981
6.	*Theobroma cacao* (Cocoa)	26	Hor *et al.*, 1984
	T. *cacao* cv. Forastero	22	Vanitha, 2001
7.	*Litchi sinensis* (Litchi)	40	Fu *et al.*, 1990

Contd...

8. *Mangifera indica L (Mango) Var. Alphonso*	25%	Doijoide, 1990
9. *Mangifera indica L (Mango) Var. Totapuri*	32%	Doijoide, 1990
10. *Hancornia sp.*	25	Oliveira & Valio, 1992
11. *Myristiga fragrans* (Nutmeg)	45	Sangakkara, 1993
	20	Madusudanan and Babu, 1994
12. *Artocarpus heterophyllus* (Jack)	36	Shylla Merlin & Palanisamy, 2000
	31	Chandel *et al.*, 1995
13. *Euterpe edulis*	39	Martins *et al.*, 2000
14. *Persia americana*	49	Raja *et al.*, 2001
15. *Murraya koeniigii* (Curry leaf)	34	Raja *et al.*, 2001(a)
16. *Areca catechu* (Betelnut)	33	Raja *et al.*, 2001(b)
17. *Eugenia jambolana* (Jamun)	20.2	Srimathi *et al.*, 2001
18. *Archontophoenix alexandrae*	31.5	Cibele *et al.*, 2003
19. *Nothadpodytes nimmoniana*	15.4	Radha *et al.*, 2010

Reports for critical moisture content vary among researchers for the same crop. In case of jackfruit seeds, while Shylla Merlin and Palanisamy (2000) reported it as 36%, Chandel *et al.* (1995) as 31%, Krishnasamy (1990) as 39% and Bhattacharyya and Basu (1992) as 27%, this huge variation may be due to mode of collection, delay in initiation of experiment or other factors making it difficult for a practitioner of seed storage to make decisions. But it is this very nature of recalcitrant seeds that posed huge challenge in preserving them.

Fruit maturity has a profound influence on seed recalcitrance behaviour. Chandel *et al.* (1995) found that the moisture content of seed reduced as they mature meaning that moisture content does decrease but not substantially as that of orthodox seeds (Table 2.4.). Chandel *et al.* (1995) further demonstrated that the lower level moisture content for 50% viability retention was also variable with immature seeds being more susceptible at higher moisture content. This holds good for orthodox seed. But when embryonic axis was concerned, different picture emerged and crops varied for their lower moisture limit for 50% viability as seen in the Table 2.4. Tea as a seed could not withstand loss of moisture beyond 24.3% in contrast, embryonic axes could be dried to 7.4%. While cocoa seeds as whole could not stand desiccation beyond 35.2% the embryo could not withstand loss beyond 55%. Such contrast needs to be explained.

It was further realized that, the moisture content did not reduce considerably in recalcitrant seeds as it does for orthodox seeds. Though the moisture reduction was same until the seeds were partially mature, but as the seed matures, the reduction in moisture was significant in orthodox seeds but not so in case of recalcitrant seed. Similar situation was noticed among embryonic axes extracted from such seeds (Table 2.6.). The variation of sensitivity could be attributed to the amount of non-freezable water present at various stages of maturity.

Table 2.6 : Comparison of moisture content of seed and embryonic axes

Crop	Immature	Partially mature	Fully mature
A. Moisture content of seed (%) fresh weight basis			
Tea	93.3	80.1	40.2
Cocoa	90.1	90.5	41.3
Jack fruit	90.3	85.2	50.1
French bean	53.1	43.3	11.2
B. Embryonic axes			
Tea	94.2	60.4	50.3
Cocoa	91.4	76.4	64.4
Jack fruit	64.3	74.6	67.2
French bean	61.4	44.5	12.0

Research has further proved that desiccation sensitivity is not the same in all parts of the seed. Moisture content of desiccation sensitive seeds varies within the seed and its components. Evaluating the moisture content of seeds and their components of seven species viz., *Cordyla africana, Enchephalartos natalensis, Encephalartos gratus, Dysoxylum spectabile, Strychnos spinosa, Strychnos madagascariensis* it was found that except for S. *spinosa*, the axis water content was higher than the storage tissues (Deon and Craig, 2006) (Table 2.7.). This is one of the chief characteristics of recalcitrant seed. The high embryonic axis moisture means, higher metabolic activity and thus faster degradation of tissues and eventual death. C. *africana*, E. *natalensis* and E. *gratus* showed considerable decline in viability with even megre desiccation and S. *spinosa* retained viability at lower moisture content. It was concluded that estimation of moisture content for the whole seed in case of recalcitrant seed would undermine the desiccation sensitivity of embryonic / germinative axis.

Table 2.7 : Water content of whole seed in comparison to seed parts (grams of water per gram dry mass)

Species	Seed mass (g)	Whole seed	Water content $g.g^{-1}$ storage tissue	Axis
Cordyla africana	± 10*		1.04 (±0.11)	1.72 (±0.14)
Encephalartos natalensis	6.98 (±0.41)		0.58 (±0.05)	1.01 (±0.07)
Encephalartos gratus	6.40 (±0.42)		0.51 (±0.01)	1.49 (±0.09)
Dysoxylum spectabile	2.21 (±0.19)	1.06 (±0.04)	0.76 (±0.04)	2.23 (±1.21)
Strychnos spinosa	1.29 (±0.14)	0.63 (±0.01)	0.63 (±0.01)	0.33 (±0.20)
Strychnos magascariensis	1.14 (±0.15)	0.78 (±0.14)	0.78 (±0.14)	1.63 (±0.40)
Bruguiera gymnorrhiza	12.58-20.88 (±3.29)	1.57 (±0.18)	1.77 (±0.10)	1.97 (±0.08)

Intra-varietal differences in critical moisture have also been found in number of species, eg., the critical moisture content for mango varieties are 25% in Alphonso, 32% in Totapuri (Doijode, 1990), 28% in Neelum and 34 % in Goa (Girija, 1998).

The extent of damage that occurs on the removal of water from desiccation sensitive seeds and thus the water content at which they lost viability, depends upon number of factors like the rate at which the seeds are dried and their metabolic activity at the time they subjected to drying (which affects the 'metabolism-induced damage') as well as the extent to which any of the protective mechanisms are expressed (Pammenter and Berjak, 1999). Pritchard and Manger (1998) also stated that the truly recalcitrant seeds can't survive the removal of any structure-associated water. Drying of desiccation sensitive seeds induced the damage in cell membrane that could not be revived after imbibition and finally results in loss of viability (Kundu and Kachari, 2000). Sudden loss of water, gradual deterioration of cellular organization, breakdown of cellular storage components (Farrant *et al.*, 1988), depletion of food reserves and accumulation of toxic substances (Copeland and McDonald, 1995) are some of the causes of desiccation induced seed deterioration.

Faria *et al.* (2005) while researching with desiccation tolerance of germinating *Medicago truncatula* seeds discovered that, the most susceptible stage was when the radicle elongated beyond 2 mm. Because upto radicle enlargement of 2 mm, there was no DNA synthesis or cell division, beyond 3 mm, the process of cell division has been initiated and desiccation tolerance was lost due disruption of microtubles and tubulin granules. This damage could be reduced by subjecting the seeds to polyethylene glycol treatment, which upon rehydration after dehydration were able to reconstitute microtubular cytoskeleton and continue normal development. Thus, it is possible to reduce desiccation induced damage by PEG treatment. This technique can be used for preservation of recalcitrant embryonic axes.

Seed moisture plays an important role at structural level, cellular level and imparts turgor pressure by filling the intercellular spaces and provides hydrophilic and hydrophobic association at molecular level, it also controls intermolecular distances that determine conformation of proteins and polar lipids, partitioning of molecules within organelles and reactivity among metal ions. Water also plays a role in controlling metabolism as a reactant and product of many reactions. As a diluent, it affects the chemical potential of other molecules, potentially shifting the likelihood of reactions. Water also provides a fluid matrix that allows diffusion of substances to reactive sites. Hence recalcitrant seeds that are highly metabolically active even when mature, will not tolerate loss of moisture.

2.3. Dessication Sensitivity and Drying Rate

Recalcitrant seeds are highly susceptible to loss of moisture thus reducing their storage potential. Seeds of mango were viable for just 4-5 weeks, jackfruit seeds for 15 days, phalsa (*Grewia asiatica* L) for 90-100 days, mangosteen for 5-7 days, nutmeg (7 days) coffee (3 months) under ambient condition (Srimathi and Sastri, 2004), cocoa (4-7 days) (Ponnusamy *et al.*, 2007).

Desiccation sensitivity is affected by drying rate and temperature. Slow drying conditions help plants to stay under intermediate water content for longer period, allowing damage associated with metabolic activity to accumulate. While faster drying conditions allow the tissues to pass through the intermediate water range rapidly, reducing damage accumulation. Hence, excised embryos or embryonic axes of desiccation sensitive seeds can survive to lower moisture conditions when dried rapidly than when dried slowly (Berjak *et al.*, 1990).

Some researchers have attempted to dry the seeds either using oven (40°C) (fast drying) or over silica gel (25°C) (slow drying), while other researchers have used saturated salt solutions. These methods have their own advantages and their disadvantages. Oven drying was at several levels not very useful. While silica gel needs long periods of drying and saturated salt solutions may be harmful. Delgado and Barbedo (2007) reported that both oven drying and silica gel desiccation methods did not help in retaining viability beyond 10%.

Seed composition reflected on desiccation sensitivity. Among the two sub genuses of Oak tried, genus *Erythrobalanus* can be dried safely to 30% while genus *Lepidobalanus* needs higher moisture viz., 45-50%. This variation was due to seed composition (Bonner, 1996). *Erythobalanus* seeds stored fat while *Lepidobalanus* reserved carbohydrates, hence needed higher storage moisture content.

2.4. Chilling Sensitivity

We can store the orthodox seeds even up to temperatures of -196°C without loss of viability. But the recalcitrant seeds cannot be stored at lower temperatures; some species do not survive even the temperatures of 10-15°C. It is because of chilling injury to the seeds, which varies according to species. Stanwood (1983) has stated that there is a high moisture freezing limit (HMFL) which is the threshold and if it is exceeded, the viability of a seed sample will be reduced during Liquid Nitrogen (LN) storage. The chilling sensitivity in moist recalcitrant seeds is mainly due to the formation of ice crystals in between the cells when the moisture content is higher than 14 to 20% (Roberts, 1972). Chin and Roberts (1980) stated that even short periods below 10 to 15°C would cause loss of viability for many tropical recalcitrant species, eg., in cocoa sharp reduction in storability was observed at 15°C compared to 17°C (Hor, 1984). Jackfruit seeds stored for 22 days at 6°C germinate well (80-86%) but none after 38 days (Panggabean, 1979).

The seeds of many tropical species are killed at sub-ambient temperatures. Species like *Theobroma cacao* (Hor *et al.*, 1984), *Nephelium lappaceum* (Chin, 1975), *Dryobalanops aromatica* (Jensen, 1971), *Hopea odorata* (Tang and Tamari, 1973), *Shorea ovalis* (Sasaki, 1976), *Myristica fragrans* (Madusudanan and Babu, 1994) and *Garcinia mangostana* (Winters, 1953) were very sensitive to chilling temperatures. Tang and Fu (1993) reported low viability of jackfruit seeds stored at 5-10°C after 40 days due to chilling injury. Song *et al.* (1984) found that at 5°C almost all seeds were dead after 6 months, but at 15 to 20°C no loss of viability had occurred in *Hopea hainanensis*. Similarly, Tompsett (1985) found sudden loss of germination ability at 6°C than at 11°C in *Shorea robusta*. Nutmeg seeds lost viability when subjected to 5-8°C storage. In mango, 3°C was lethal and Alphonso cultivar was dead at 8°C (Srimathi and Sastri, 2004).

In case of *Calophyllum brasiliensis,* the seeds are very sensitive to 5°C but best results are achieved for seeds with a moisture content around 30% at ambient temperature and 15°C (Vasquez, 2001). Rajeswari Dayal and Kaveriappa (2000) found that the *Hopea parviflora* and *H. ponga* seeds subjected to freezing (0 ± 2°C) showed a rapid decline in the subsequent germination percentage at 30 ± 2°C, as well as vigour, even though the moisture content of these seeds did not vary much. In *Hopea,* Tamari (1976) observed a decline in the development of radicle with decrease in temperature. The reason might be due to the decreased metabolic activity of the seedlings at lower temperatures.

This sudden loss of germination was due to chilling injury to the seeds at lower temperatures (King and Roberts, 1979). Tompsett (1992) also reviewed that the reduced germination following exposure to temperatures in the range 0 to 16°C (chilling damage) occurs in all moist Dipterocarp seeds.

Seeds of Tea, cocoa and jackfruit were able to withstand drying upto 24, 35, 31% moisture respectively but could not tolerate immersion in LN (Chandel *et al.,* 1995). Some extent of cryopreservation of partially or fully mature embryonic axes of tea and jackfruit could be obtained when dried to 14% moisture while cocoa axes were totally sensitive at all stages of physiological maturity.

2.5. Cellular Level Characteristics of Recalcitrant Seed

Apart from possessing larger sized seeds, recalcitrant seeds can also be differentiated from other seeds to some physical attributes at cellular level (Berjak and Pammenter, 1996, 2000).

2.5.1. Anatomical changes occurring in recalcitrant seeds

The desiccation of recalcitrant seeds to below the critical moisture content may result in the cellular and sub-cellular changes particularly cell wall rupturing and cytoplasmic fragmentation (Chin *et al.,* 1984).

The sub-cellular deterioration accompanies with desiccation was reported by several workers in many recalcitrant seed crops like *Avicennia marina* (Farrant *et al.*, 1988), rubber (Normah and Chin, 1989), cocoa (Ruhl and Dambroth, 1989), *Zizania palustris* (Berjak *et al.*, 1994), and *Ekebergia capensis* (Pammenter *et al.*, 1998). Chin *et al.* (1981) observed that the nucleus, which was irregularly shaped and sometimes the membranes surrounding the nucleus were hardly recognizable and a nucleolus was missing in dried rubber cells. Similarly, Ruhl and Dambroth (1989) observed complete disintegration of the cell constituents in desiccated cocoa seeds while Raja (2001) encountered similar experiences with arecanut seeds (Figure 2.1) The ergastic substances are metabolites such as fats, starch or crystals, which are the passive products of cellular activity. The differential ergastic build up in the necrotic cells of the dried samples had been attributed to the result of deranged metabolic processes in the plant system. The highly vacuolated cell structure of the seed may lead to a situation where the vacuoles lose their turgidity (Pammenter *et al.*, 1994). It is also possible that the water requirement for the developing embryonic axes may be met by the withdrawal of water from the other tissues in the seed. Such a withdrawal of water from the vacuolated cells contributes to water stress in these cells, which ultimately leads to cellular damage (Thompsett, 1984 and Berjak, 1989).

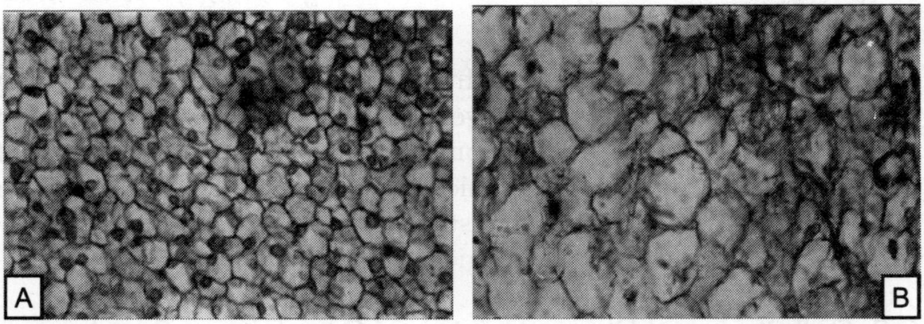

Figure 2.1 : Microtome section of arecanut embryo (87µ) showing A. Normal tissue, B. Desiccated tissue with complete cellular disruption.

2.5.2. Role of vacuoles

One of the major attributes of recalcitrant seeds was their ability to withstand mechanical stress. Iljin (1957) attributed this quality to the vacuole volume reduction either by shrinkage of intercellular spaces by these organelles due to accumulation of solutes or insoluble reserve material that provides the ability of a recalcitrant seed to absorb mechanical shock. This was further corroborated by Farrant *et al.* (1997) in *Avicennia marina, Aesculus hippocastanum* and *Phaseolus vulgaris*. Their findings had far reaching consequences in ferreting out the quality of recalcitrant seeds. *Avicennia* seeds are shed without dehydration and can't survive even lower than 0.5 g/g (33% wwb). It was found that vacuoles occupy more than 60% of the cell volume in all embryonic axis and more than 90% in cotyledonary cells and significantly the vacuoles in these tissues contain less insoluble reserves. While seeds of *Aesculus* though loose moisture and achieve desiccation tolerance to a range of 0.42 to 0.25 g/g. Vacuoles at this stage are very less in both embryonic axis and cotyledonary tissues. These vacuoles contain large starch filled plastids and protein bodies. Seeds of *Phaseolus* that are orthodox, have very low vacuolar space. The vacuolar cells in cotyledon accumulate an amorphous, insoluble substance. Thus it is the vacuole that is the key to desiccation tolerance with the recalcitrant seeds possessing large vacuoles filled with soluble materials while orthodox seeds possess minute vacuoles with more of insoluble deposits preventing lethal mechanical damage caused upon loss of moisture (Farrant *et al.,* 1997).

2.5.3. Damages to cytoskeleton

The cytoskeleton helps to form the cell shape and in organization of the cytoplasm and nucleus. These are mostly composed of microtubules and microfilaments. Microtubules consist of polymerized tubulin and microfilaments that are composed of F actin, a copolymer of G actin. In oak seeds, it was found that an extensive microfilamentous network was found at the root tip that gets dismantled upon dehydration. In orthodox seeds, the reassembly of micro skeletion takes place. However, in *Quercus*, extreme dehydration does not lead to reassembly (Mycock *et al.,* 2000). This attribute may be one of the reasons for the recalcitrant seeds not being able to withstand drying. Further, certain cytomatrical enzyme systems exist as multienzyme particles in plant cells (Hrazdina and Jensen, 1992) that act as key or anchor enzymes to microfilaments of cytoskeleton. These particles would be destroyed at low temperature thus preventing root primordial initiation.

2.5.4. Integrity of nucleolus

In orthodox seeds, desiccation does not lead to damage to DNA and its repair is rapid upon rehydration. Since DNA has different conformational states depending on seed moisture content, which makes it extremely difficult to predict its role in desiccation (Osborne and Boubriak, 1994). It was found that DNA helix had an increased number of base pairs per turn upon dehydration. It was also suggested that, protein glycation can increase the incidence of DNA conformation upon dehydration or even the possibility of methylation of cytosine that could favour the Z form of DNA.

Apart from changes in DNA shape, changes in chromatin should also simultaneously occur for stabilization by shrinking upon dehydration as postulated by (Sargent *et al.*,1981). This state is conserved upon redhyration and is reversed during germination when desiccation sensitivity ensues (Deltour, 1985). Ivanov and Zlatonova (1989) suggested that the chromatin stabilization in dry orthodox seeds due to change in H1-histone : nucleosome ratio from 2:1 to 1:1 could be the cause for desiccation tolerance.

Moreno Diaz de la Espina (1995) postulated that filaments found in nucleo skeletion especially, lamins that form a complex with matrix found beneath the nucleo skeleton was supposed to stabilize and fix the chromatin in distinct domain. Upon rehydration, it was possible for the chromatin to resume shape and functionality leading to radicle protrusion. It was not possible to revive desiccation tolerance in germinated seeds where radicle protrusion was beyond 2 mm in *Medicago truncatula* seeds due to disassembly of micro tubules (Jose Faria *et al.,* 2005). PEG treatment could restore desiccation tolerance to some extent beyond which the cytoskeletal structure was damaged completely.

Hence, it may be presumed though the role of DNA, chromatin and microtubles are not clear, their ability to withstand desiccation and reconstitution upon rehydration is however essential for the regeneration of cells in seeds.

2.5.5. De-differentiation of intercellular components

Orthodox seeds are able to de-differentiate cellular components upon drying. The cellular structures are simplified and minimized. However upon rehydration, this process is reversed and the cells attain their sophistication that is necessary for further growth and development. In desiccation sensitive seeds like *Avicennia*, the extent of mitochondria was the highest and substantially less for *Aesculus hipoocastanum*. In *Phaseolus*, this was very less even preceding the onset of drying. The embryonic axis in both the recalcitrant seeds also had significantly higher proportion of mitochondria. The cristae were also well formed as in an active cell. While, in *Phaseolus*, the mitochondria were completely de-differentiated even at tissue water contents compared to recalcitrant seeds (Farrant *et al.*, 1997). Hence, it can be seen that the de-differentiation of major organelles is a major factor for desiccation tolerance in orthodox seeds. Farrant *et al.* (1997) indicated that, mitochondrial revival upon imbibition follows shrinking prior to maturation drying.

Further, studies have revealed that respiration occurs even when the moisture content is at 20% and when the moisture content increases, the metabolism rate gallops leading to stupendous generation of free radical (Vertucci and Farrant, 1995). However, orthodox seeds are able to tolerate this during maturation drying. This ability to tolerate drying could be attributed to the efficient methods of switching off of metabolism in desiccation tolerant seeds as seen by de-differentiation of mitochondria in *Phaseolus* prior to maturation drying. While in desiccation sensitive seeds, potential lethal damage is found when water content is in the range of 0.45 - 0.25 g per g of seed. Loss of viability is slower when water is lost slowly, but rapid dehydration improves the possibility of survival at lower moisture contents thus marshalling the idea of drying embryonic axis quickly to facilitate easy storage. Embryonic axis did not survive flash drying for more than two days but can be stored under cryogenic conditions. As the embryonic axis are metabolically active, flash drying may arrest this metabolic activity thus preventing further damage by free radicals.

2.5.6. Continuation of cell multiplication

In recalcitrant seeds, cells in the embryonic axis are continuously in division. In *Avicennia mariana*, the DNA replication in root primordia is curbed only for a day or so at which time, the water is lost by 16-18%, after which it falls into the water and resumes it growth and germination. Comparably in orthodox seeds, the embryonic axis consists of cells in haploid condition and during cell cycle, of the four distinct phases *i.e.*, G1 phase which is the resting phase for the cell, followed by S phase wherein DNA replication occurs which is followed by G2 phase where the DNA is doubled and again followed by G2M phase when mitosis occurs and reduces the doubled DNA content back to its original content. Among these four distinct phases, Brunori (1967) found that seeds of *Vicia faba* contained cells in G1 phase caused by cessation of DNA replication. Thus, the conservation of DNA occurred by curtailing cell division, which was not seen in recalcitrant seeds. Finch Savage *et al.* (1998) observed in *Acer pseudoplatanus* that cell cycling was arrested with embryonic cells in G1 phase. In *Azadirachta indica,* the 2C DNA level was found to occur and no 4C DNA state was recorded. Thus, the DNA replication states vary from species to species, which high level of 4C DNA in Avicennia and to a 60:40:4CDNA:2CDNA level in Acer to completely 2C DNA in *Azadirachta.* No clear idea emerged by studying the status of the cellular DNA levels.

2.5.7. Biochemical changes during desiccation

The dehydration-induced deterioration of the cell membrane in recalcitrant seeds is indicated by high increase in leakage of solutes (Chin and Roberts, 1980). Heydecker (1972) suggested that weakening of cell membrane might be the cause of leaching metabolite like electrolytes and other cell soluble compounds into the imbibing medium. This loss in membrane integrity results in the enhanced efflux of cellular constituents like sugar, amino acid, phenol and phosphates (Nautiyal and Purohit, 1985). Also the membrane conductance depends upon the concentration of the carrier for electrolytes in the membrane, the density of the membrane surface charge, the quantity of permanent low and the ionic strength on both the sides of the membrane (McLanghlin *et al.,* 1970).

It was observed that the desiccation of developing seeds was characterized by the accumulation of a particular set of m RNA and related proteins called 'Late Embryogenic Abundant (LEA) proteins' in the desiccated state (Blackman *et al.,* 1991). Further, Farrant *et al.* (1993) reported that *Avicennia marina* seeds do not produce LEA proteins, which supported the suggestion that production of such proteins might facilitate desiccation tolerance. The group of LEA's that has received the most attention is the LEA D_{11} family, also known as the "dehydrins", which were responsible for dehydration stress and in recalcitrant seeds appears to be anomalous (Pammenter and Berjak, 1999). The accumulation of a class of LEA proteins (Close *et al.,* 1989; Dure *et al.,* 1989; Dure, 1993), often called 'dehydrins' during development is to protect against desiccation damage, particularly to membranes. Desiccation sensitivity in seeds has therefore been suggested to result from the absence of dehydrin proteins (Bradford and Chandler, 1992; Farrant *et al.,* 1992). Farrant *et al.* (1992) reported that no such new proteins are produced during the late stages of development in the highly desiccation - sensitive seeds of *Avicennia marina.*

Phenols are aromatic compounds which include an array of components like tannins, flavonoids etc. The seeds of many tropical plants contain high concentrations of phenolic compounds and phenolic oxidases. These compounds are normally compartmentalized within cells. On desiccation, the cell membranes are damaged and the phenolic compounds are released. They are then oxidized and protein - phenol complexes are formed leading to loss of enzyme activity (Loomis and Battaile, 1966). In arecanut seeds reduction in phenol content during desiccation was evident due to the destruction of the tanniferous cells and release of phenolic acids to the surrounding cells.

The fat content of recalcitrant seeds declines with steady increase in free fatty acids due to desiccation of seeds (Raja *et al.*, 2002). Clatterbuck and Bonner (1985) observed a steady decrease in crude fats of *Quercus robur*. Georgi *et al.* (1983) found that the untreated rubber seeds deteriorated rapidly in storage with a considerable development of acidity in the oil. The oxidation of fat contributed to the accumulation of free fatty acids might be the reason. Accumulation of free fatty acids in seeds is subjected to slow and constant attack by oxygen resulting in the production of hydrogen peroxide and other oxygenated free fatty acids and free radicals. Due to the accumulated oxygenated fatty acids, death occurs through the destruction of the respiratory pathway by the toxic aldehydes and free radicals formed by peroxide and epoxide decomposition (Harrington, 1973). Koostra and Harrington (1969) stated the Millard reaction in which the degradation of lipo-protein cell membranes by free radical induced lipid peroxidation reaction has been suggested to be a basic reason of senescence and ageing. Free radical activity has been associated with viability loss in several recalcitrant seeds during desiccation (Hendry *et al.*, 1992; Finch - Savage *et al.*, 1994; Chaitanya and Naithani, 1994). Various authors (Simon, 1974; Senaratna and McKersie, 1986; Bhattacharyya and Basu, 1992; Li and Sun, 1999; Raja *et al.*, 2002) have also supported that lipid peroxidation is a cause of membrane deterioration during desiccation and that the damage would result in loss of semi-permeability, increase in free fatty acids and accumulation of thiobarbituric acid - reactive substances such as melondialdehyde.

The other cause for viability decline in recalcitrant seeds is the production of free radicals. Results obtained by Senaratna *et al.* (1987) with microsomal membranes seem to show that free radicals induce de-esterification of membrane phospholipids rather than change in fatty acid saturation. But according to Leprince *et al.* (1994)

mitochondria might be the primary source of electrons leading to the production of stable free radicals in desiccation intolerant radicles.

Volatile aldehydes constitute an important component of the gaseous emanations of deteriorating seeds (Harman *et al.*, 1982). Besides volatile aldehydes, other growth inhibiting substances in the gaseous emanations were also observed by Woodstock and Taylorson (1981). Bailey *et al.* (1962) found various volatile compounds such as isovaleraldehyde, isobutyraldehyde, propionaldehyde, methanol, acetaldehyde, methyl acetate and diacetyl in cocoa bean seeds. Bhattacharyya and Basu (1992) has found that vigour bioassay of gaseous emanation of jackfruit seeds and demonstrated that deteriorated seeds produced larger quantities of volatile growth inhibitory substances during germination. Desiccation of tea, cocoa and jackfruit seeds lead to increased leachate conductivity, lipid peroxidation products and soluble carbohydrates and damage to cell membrane when freezing and thawing occurred as revealed by ultrastructural changes (Chandel *et al.*, 1995, Raja *et al.*, 2002).

2.5.8. Impaired protective mechanism in recalcitrant seeds

Generally in seeds, the multiple protective mechanisms exist against the highly reactive oxygen radicals. They involve free radical and peroxide-scavenging enzymes such as catalase, peroxidase and superoxide dismutase. Recalcitrant seeds (or their embryos) do appear to possess antioxidant mechanisms (Hendry *et al.*, 1992). However, these protective mechanisms may become impaired under conditions of water stress (Smith and Berjak, 1995) or they are ineffectual in terms of protecting against desiccation damage. During desiccation of arecanut seeds the activities of scavenging enzymes such as catalase and peroxidase were reported to decrease and the cells loose the protective mechanism of tolerance to desiccation (Raja *et al.*, 2002). This indicates that oxygen free radicals are continuously produced in the plant system. When exposed to stress they produce the enzymes in large amounts to eliminate the free radicals. As they exceed the eliminating ability, the plant system is damaged. The peroxidase enzyme is also involved in the dehydrogenation of a large number of organic compounds like phenols and aromatic amines. The destruction of the enzyme can lead to the accumulation of toxic substance in the seed. Hendry *et al.* (1992) found a rapid accumulation of free radicals accompanied by decreasing activity of antioxidant enzymes and declining α-tocopherol content.

In case of *Quercus robur* and Q. *borealis* a very high amylolytic enzyme activity was noticed at the beginning of storage. However, it was quickly declined to non-detectable levels by 8 months (Szezotka, 1975). Polyphenol oxidase, another group of enzyme capable of degrading hydrogen peroxide and using it as an electron acceptor is capable of catalyzing the same type of reactions as peroxidase (Kahn, 1983). Nevertheless, some recalcitrant seeds have retained active metabolism and thus showed an enhanced level of polyphenol oxidase activity during initial drying period. However the suppression of polyphenol oxidase activity during later desiccation was noticed with loss of seed viability (Nkang *et al.*, 2000; Raja *et al.*, 2002).

The structured water in the seed is involved in ensuring the precise functioning of these multi-enzyme systems (Clegg, 1979). All metabolic activities probably do, take place in structured water. Loss of structured water results in the disruption of metabolism. In case of orthodox seeds, this presumably does not occur, as shown by their tolerance to desiccation. But in recalcitrant seeds, the situation is quite different (Berjak *et al.*, 1984).

2.5.9. Putative role of Late Embryogenesis Abundant (LEA) proteins

Corn seeds produce some proteins that occur only during developmental phases when there is desiccation (Bochicchio *et al.*, 1988). Likewise, soybean seed produces polypeptides related to water stress (Rosenberg & Rinne, 1988). These polypeptides are like LEA protein found in cotton seeds (Galau *et al.*, 1986) and *Ranunculus sceleratus* seeds (Wechsberg *et al.*, 1993). Similarly some protein form in barley seed is enhanced during the desiccation process that imparts desiccation tolerance (Bartels *et al.*, 1988).

However, absence of proteins related with desiccation was not the only unique factor for the lack of desiccation tolerance of recalcitrant seeds as indicated by experiments performed by Blackman *et al.* (1991 and 1992), Bradford & Chandler (1992) and Gee *et al.* (1994), who opined that it ranged with some seeds tolerant and others intolerant to desiccation. Seeds intolerant to desiccation which have desiccation-protein like orthodox seeds, do not survive after drying. It is thought that these LEA proteins have fundamental function in tolerance desiccation process.

2.5.10. Role of oligosaccharides in desiccation sensitivity

Some sugars like sucrose and other oligosaccharides are present in desiccation tolerance process in corn, soybean and pea seeds during the beginning of germination (Koster & Leopold, 1988). After these oligosaccharides were removed, the desiccation tolerance disappears. Corn seeds during germination have high raffinose and sucrose contents and no glucose and fructose. The opposite occurs during the development of the desiccation intolerance (Leprince *et al.*, 1992); the same being observed in alfalfa and carrot somatic embryos (Tetteroo *et al.*, 1994) and in cauliflower seeds (Hoekstra *et al.*, 1994). The ability of desiccation tolerance in some seeds may originate from mechanisms of avoidance of cellular destruction during water losses. When a non-tolerant desiccation cell dehydrates, its solutes can be more concentrated (probably increasing the destructive chemical reactions), some solutes can crystallize (modifying ionic resistance and pH intracellular solution), proteins begin denaturing and the membranes rupture (with compartmentalization losses). The presence of great amounts of soluble sugars within a cell can prevent from harmful damages of desiccation building up hydrogen links and then, substituting water in preserving the water structure orientation, when hydrated (Koster, 1991). This ability probably derives from the capacity to reach glassy state at room temperature (Koster, 1991). This state depends on the ratio of sucrose : raffinose, especially to prevent of crystallization. The combination among sugars present in desiccation tolerance cells affords glassy state reaching at temperatures higher than 0°C and lower than 0°C in non-desiccation tolerant ones. The glassy state of water, its high viscosity, would delay solutes crystallization and total dehydration. Molecular mobility inhibition prevents denaturation and undesirable biochemical reactions, guaranteeing the survival of desiccated seeds (Williams and Leopold, 1989).

□□□

Chapter 3

Collection and Extraction of Recalcitrant Seeds

Due to their poor storability, collection and cleaning, proper storage of recalcitrant seeds becomes essential. Proper collection will envisage proper transport, processing and storage. This becomes all the more important for *ex situ* conservation. Bonner (1996) provided a comprehensive list of decisions to be taken while handling and transporting recalcitrant seeds. Seeds upon harvest from trees must be processed on-site. The decision whether to extract the seed from the fruit at a central seed processing depot or close to the site or soon after, must be made in the light of local conditions. Reduction in bulk and weight of the collected material greatly facilitates transport and for this reason the extraction of the seed from relatively bulky fruits may be desirable at an early stage, even if the more critical operations of dewinging, cleaning and final drying are better done at the central depot.

3.1. Identification of Candidate Trees and Phonological Studies

Srimathi (1997) reported that identification of candidate plus tree and age of the tree was important in Jamun (*Syzygium cuminii*) as significant variation was found for seed weight and seed yield among trees. Fruit bearing was also irregular. Twenty year old trees were more suited. Fruits collected at bottom of crown and middle of branches and from north side of tree had higher quality. Seeds collected from Bhavanisagar (Tamil Nadu) faired better than from Coimbatore (Tamil Nadu). Seeds collected from crown or immediately after falling on the ground had better quality. There is 65% flower drop in first five weeks, 19-21% flowers/fruits drop prior to maturity and 12-15% alone reach maturity, under scoring the need for planning the harvest. Neem drupes can be collected from 96th day onwards. The weight of 1000 drupes at physiological maturity was 2950 g, 248 g and 146 g respectively and the number ranged from 3950 to 4000 per kilogram. Drupes at greenish yellow stage collected from tree crown were best. Change for colour from green to yellow was the physiological maturity index. Trees approximately 20 years old are the best. Western zone of Tamil Nadu had higher germination, vigor, low free fatty acids and high protein (Bharathi,1999).

3.2. Harvesting and Collection of Fruits or Seeds

Fruits must be harvested at maturity or when they are about to be shed. However it is difficult for most forest tree species and viviparous. A thorough knowledge of the flowering or masting behaviour of the species is essential. This will help in logistics essential for programming a seed recovery strategy. Since it is not correct to collect the seeds fallen on the ground as they may be over ripe, infected or damaged by granivours or frugivorus predators, one must attempt to collect the fruits/ seeds from tree itself. This may help in collecting the seeds without damage and help us determine the exact nature of seed in question. Monitoring the moisture content and dry weight of seeds during their development can be useful in deciding when to harvest. In case of trees and shrubs, seeds must be collected from single plant. If the sample is insufficient, then seeds at the same maturity stage and age must be collected. Seeds collected from different locations and of different maturity must not be mixed. Most tropical evergreen forest trees are huge and grow to a tremendous height. Such trees need specialist climbers.

Working with sycamore seeds, Hong and Ellis (1990) reported that seeds collected during July had high moisture and low viability quotient while those collected during August to September had higher germination thus underlining the need to assess the fruiting and maturity behaviour before seed collection (Figure 3.1.).

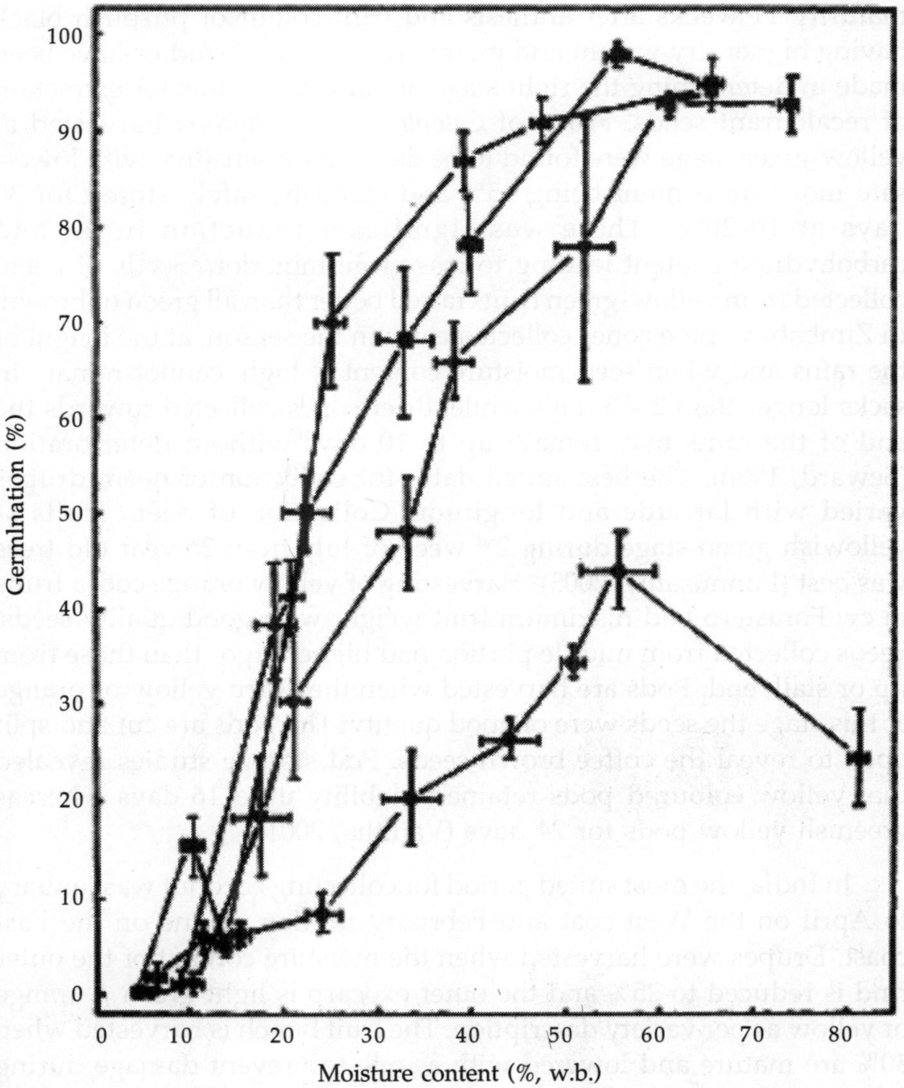

Figure 3.1 : Relationship between germination (%), moisture content of dried Sycamore seeds (*Acer pseudoplatanus*) harvested during different dates (12th July (•), 16th August (▼), 20th September (■) or 18th October (▲)).

Courtesy: Hong and Ellis, (1996)

Attainment of physiological maturity wherein maximum growth and development has been achieved is purported to be the right stage for harvest. In Jamun, flowering season was June, fruiting season was July-August and fruit set was only 6.0%. Fruits reached physiological maturity 11 weeks after anthesis and fruit colour of purplish black having higher dry weight and germination. Several studies have been made in determining the right stage of harvest of fruits for extraction of recalcitrant seeds. Fruits of *Calophyllum inophyllum* harvested at yellow-green stage were found to be desiccation sensitive with lowest safe moisture content being 15% and could be safely stored for 30 days at 10-20°C. There was significant reduction in oil and carbohydrate content leading to loss of germination. Seeds of neem collected from yellow-green fruits faired better than all green or brown. In Zimbabwe, pine cones collected early in the season, at the height of the rains and when seed moisture content is high, cannot remain in sacks longer than 2 - 3 days while, drier cones collected towards the end of the rains may remain up to 10 days without deterioration (Seward, 1980). The best suited dates for collection of neem drupes varied with latitude and longitude. Collection of neem fruits at yellowish green stage during 2nd week of July from 25 year old trees was best (Ponnusamy, 2003). Harvesting of yellow orange cocoa fruits in cv. Forastero had maximum fruit weight with good quality seeds. Seeds collected from middle portion had higher vigor than those from tip or stalk end. Pods are harvested when they turn yellow or orange at this stage the seeds were of good quality. The pods are cut and split open to reveal the coffee brown seeds. Pod storage studies revealed that yellow coloured pods retained viability upto 16 days whereas greenish yellow pods for 24 days (Vanitha, 2001).

In India, the most suited period for collecting coconut was January to April on the West coat and February to May / June on the East coast. Drupes were harvested when the moisture content of the outer rind is reduced to 25% and the outer exocarp is light green / orange or yellow as per variety description. The fruit bunch is harvested when 50% are mature and lowered with a rope to prevent damage during fall. Big sized nuts had maximum endosperm and retained high percentage of viability. Nuts that are light, under sized, cracked or uniform were rejected. Ripe fruits were allowed to shade dry for a month when the rind looses moisture and becomes light and dry before being sown in nursery.

In mango, selection must be done carefully and fruits examined for stone weevil infection. It has been found that Alphonso variety is not susceptible to stone weevil (*Sternochetus mangiferae*) infection unlike Neelum or Bangalora varieties. Mango seed stones loose viability if sown 20-30 days after extraction. Doijode (1990) found that both cultivars of mango Alphonso and Totapuri retained viability for five weeks only under ambient conditions while the moisture content was reduced from initial levels of 63.6 to 24.9% in Alphonso and 61.6 to 32.1% in Totapuri.

In Jack (*Artocarpus integerifolia*), the fruits are harvested when the outer rind shows conspicuous development of pointed protuberances and becomes greenish yellow in colour. The fruits are cut, brought and lowered gently down the tree and cut open to reveal the multiple fruits inside. The soft – sweet mesocarp is removed to reveal the brown coloured seeds. Dry and shriveled testa or cotyledons indicate loss of viability. Vivipary is noticed in early maturing types (25-30%) and late maturing types (37-39%). The seeds are dried under shade and stored in poly bags.

In avocado, trees show biennial behaviour. Being a cross pollinated crop, fruit set is poor. Further cold and heavy winds take their toll. Hence collection must be limited to healthy and bold fruits bearing bold seeds that have to be washed and immediately processed for short term storage.

Nutmeg fruits split open to reveal the seeds which shed in two to three days. Such seeds have poor viability. Hence fruits which are ripe, indicated by yellow colour of rind are collected and split open to get the seeds. The mace is then removed and seeds alone are used. The initial germination is 97% but reduced to 67% within three days (Madusudanan and Babu, 1994). Hence fruits must be collected as soon as they reach yellow colour.

Best quality acorns of Ban Oak (*Quercus leucotrichophora*) were selected by placing them in water and collecting the sinkers for further use was reported by Bhardwaj *et al.,* (2001a). Seeds of *Hopea odorata* and *Hopea parviflora* have to be dewinged for storage and transport.

Harvest of recalcitrant seeds must be rightly timed. Delay will lead to loss of vigor and may lead to precocious germination. Most of the fruits are found high in tree tops and special methods must be taken to collect undamaged fruits. Erection of nets under the tree and gently nuding them to drop can be practiced or a pole attached with wirenet at its tip can be used to collect fruits from trees as practiced for mango.

3.3. Extraction of Seeds

In some cases it has also been found that early extraction of the seed is essential to maintain maximum viability. Fleshy fruits such as the syncarp of *Chlorophora* ferment if stored in bulk for any length of time and the fermentation impairs seed viability. Same is the case of mango, jack, jamun, amla etc. Depulping of *Garcinia gumminguta* fruits was done to release seeds followed by abrasion of seeds with sand and washing with kerosene was found to improve seed handling.

Seeds of coconut (*Cocos nucifera*) are a fibrous drupe when mature. The seeds covered by a hard endocarp which has to be broken to expose the seed. The seed is covered by thin (often attached) testa with a thick endosperm and liquid endosperm. The fibrous outer rind is removed using a sharp pointed pole or crowbar. In jamun, extraction of seed after heaping for a day was better than longer durations. Similarly, Madusudanan and Babu (1994) reported that nutmeg seed has a tough shell like endocarp which has to be broken to reveal the seeds. Decoated seeds provide 100% germination with a dormancy of 6-7 days while whole seeds have a dormancy of 30-40 days. Neem drupes are washed and scrubbed over wiremesh to remove fleshy mesocarp. Neem seeds after extraction, maceration and washing, are air dired for 6 hrs and empty seeds removed by water upgrading. Sinkers thus separated were vigorous. Seeds then can be dried to 10% by sun drying or in oven at 40°C and size graded using 6.4mm sieve (Ponnusamy, 2003). Bharathi (1999) suggested neem seed scarification using concentrated hydrochloric acid at 80 ml / kilo for 1 min followed by washing with water for 4-5 times was best for getting higher germination. Mucilagenous cocoa seed extracted using H Cl @ 8 ml / 250 g pulp for 5 min recorded higher germination and vigor than other methods. Mango kernel extractor is also available, but the machine has to be monitored for getting undamaged kernels.

The use of the sun to speed extraction and drying of the seed is common in the tropics, particularly for many conifers and trees with woody, dehiscent fruits and is generally beneficial. This cannot be recommended for recalcitrant species. Rapid drying of fruits may sometimes cause them to loose viability faster than necessary. Also some times the bacterial infection or fungal infection may be exacerbated. In such cases it is necessary to keep the fruits under cover, with a good circulation of air, for one or two weeks before more rapid drying is attempted.

Species from the evergreen-wet tropical forest are easily killed by rapid drying of the fruits and they should instead be kept moist during transit (Kemp 1975).

Factors to be considered when deciding whether to extract locally or centrally include:

1. *Distance to central seed processing depot and efficiency of the transport system* : The longer the period to be spent in transit, the greater the risk of deterioration, the greater the saving in cost from transporting seeds instead of fruits and the stronger the argument for local extraction.
2. *The characteristics of the species in question* : Seeds from fleshy fruits with a high risk of fermentation, should be extracted by maceration locally and soon after collection. While dry, resistant fruits and seeds of some leguminous species can withstand a lengthy transit period under poor conditions without serious loss of viability.
3. Recalcitrant species which will not withstand drying are best despatched moist to the central seed depot as soon as possible.

In the ever-wet tropics sun drying is not practicable for collections attempted during the rainy season. Manual extraction using hand is more desirable and essential as the damage to sensitive seed coat can be minimized. Machines like blender, macerator or methods like water soaking, fermentation treatments and use of chemicals like hydrochloric acid, sodium carbonate or enzyme digestion must be avoided. Washing and cleaning of seeds using running water is generally sufficient to remove the gelatinous substance on the seed.

Even where seed extraction is done locally, it is preferable to do it where there are some facilities in buildings and communications rather than in the forest itself. In a country with a well-developed forest service, the local Forest District Head Quarters is often the best place. Seed collection in inaccessible and sparsely inhabited forests will have no such facilities and will need to improvise a seed extracting depot in the forest.

3.4. Grading of Seeds

Large seeds in clove and nutmeg and small seeds in rubber performed better in germination and vigor (Gunasekaran, 1997). In cocoa, medium sized seeds performed better and pink coloured seeds were superior to white (Vanitha, 2001). In jamun, size grading using 20/64" round perforated sieve had 28% recovery with 93% germination while in aged seeds, low vigor seeds can be separated by water floatation (Srimathi, 1997). Mangosteen seeds devoid of pulp showed improved germination of 10% than when sown with pulp (Ponnusamy *et al.*, 2007). Coffee berries of San Ramon were separated as floaters and sinkers using water. Sinkers were 93.4% and floaters 6.6%. Uniformity in maturity of berries was not noticed even after this grading. Fresh weight of berries, pericarp, bean and dry seed weights of sinkers were 9.7, 20.2, 32.8 and 72.2% higher over floaters. Pulp and parchment weight were more for floaters by 69.4 and 21.2%. The thin silvery seed coat encircling green endosperm was loosely held in floaters and tightly held in sinkers. Specific gravity separation using water thus helped in getting better quality seeds. The bean and seed weights were 60.3 and 68.2% higher over the smaller ones. In cocoa, slimy mucilage on seed coat makes extraction difficult and inhibits / delays germination and seedling growth in stored seeds. Sulphuric acid seed treatment for three minutes removed the mucilage (Ponnusamy *et al.*, 2007).

3.4. Application of Wax Coating

Friend (1964) and Wills *et al.* (1981) reported successful storage of cocoa seeds in paraffin wax coated cocoa pods. Shivarama Reddy (1987) reported similar results for coffee for 300 days with 49% viability. Raja (2001) applied hot wax over arecanut cv. Mettupalayam and stored them in gunny, polythene (350 guage) along with a portion in ambient condition. Seeds coated with paraffin wax and stored in gunny recorded a germination of 60% after 50 days coupled with higher seedling characteristics (Figure 3.2). Litchi seeds preserved in fruits coated with a combination of 0.05% Benomyl and wax emulsion (6%) and stored in polybags maintained 42% germination for 24 days (Ray and Sharma, 1987). The beneficial effect could have been the restriction of atmospheric effect on the seeds found inside the pods. However, no study on effect of storage temperature on the wax coated seed/fruit has been found, as temperature whould have profound effect on respiration of the seed/fruit. Umboh (1987) felt that the blocking of respiration due to wax application resulted in poor viability of *Shorea javanica*. Wax coating or other preservative applications have to used sparingly after thorough research.

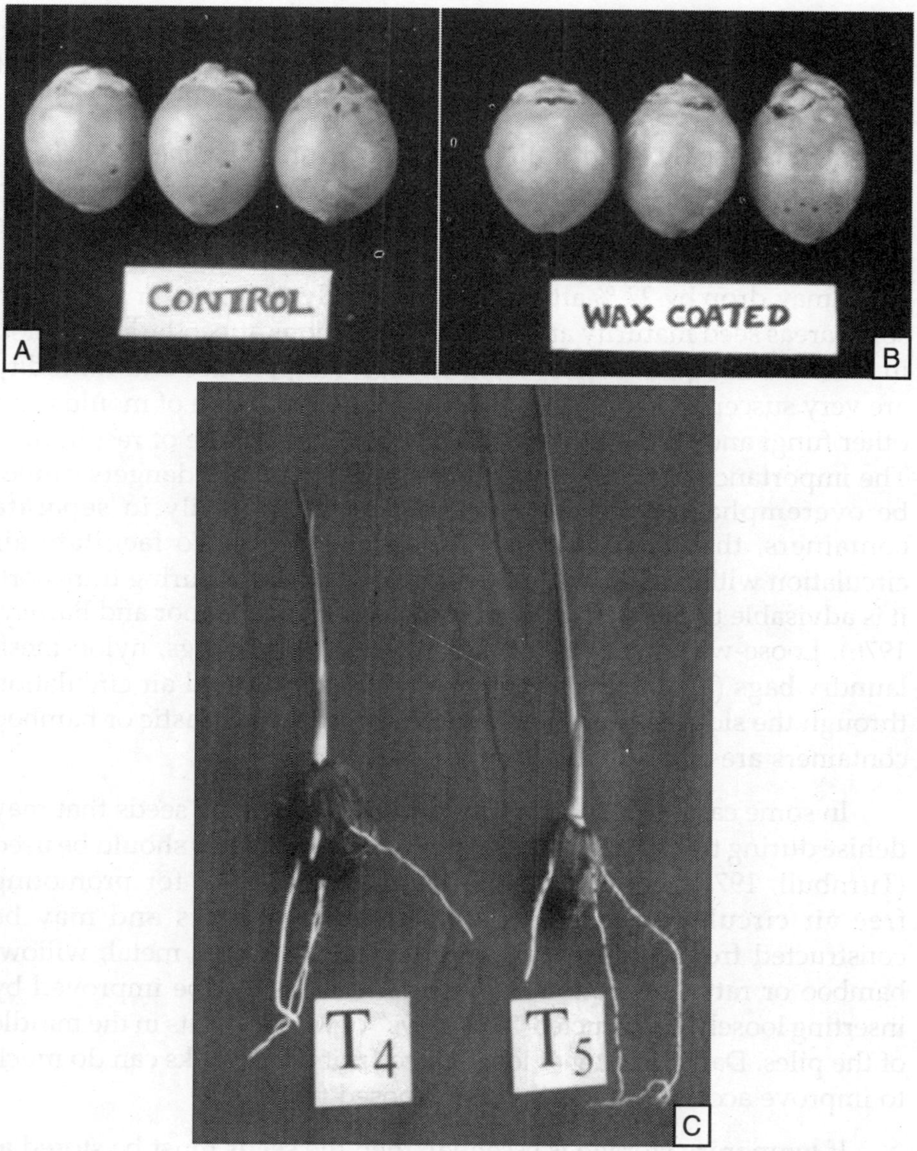

Figure 3.2 : Storage studies in arecanut. A. Untreated control, B. Wax coated C. Growth of wax coated and stored seeds after 50 days (Raja, 2001)

3.5. Transport

Depending on the nature of collection routine, seeds can be extracted *in situ* and stored or have to be transported to central lab for processing. In those cases, where the fruit has high moisture, then seeds will also possess high moisture, in such cases, these may be transported as fruits. If extracted, such seeds may germinate or loose moisture rapidly and die. If seeds are not extracted in the field, great care must be taken of the fruits both in the forest and during transport.

In large-scale seed collection operations of *Gmelina arborea* plantations in the Jari project in Brazil, great emphasis is placed on reducing the period in transit using sacks to the minimum. The average period was 1.2 days (Woessner and Mc Nabb 1979). Germination of fruits may drop by 22 % after a day and nearly to zero after a week. In some areas seed maturity and weather conditions affect the safe period in transit. Bulk quantities of fruits in high temperature and humidity are very susceptible to deterioration through the action of moulds and other fungi and through overheating due to a high rate of respiration. The importance of good ventilation in reducing these dangers cannot be overemphasized. If fruits are stored temporarily in separate containers, they should not be filled to the top. To facilitate air circulation within sacks, as well as for easy handling during transport, it is advisable to place 10-20 kg of fruits in each sack (Goor and Barney, 1976). Loose-weave hessian sacks or HDPE netted bags, nylon-mesh laundry bags (Yeatman and Nieman, 1978) allow good air circulation through the side of the container. Strong perforated plastic or bamboo containers are also suitable for transport.

In some cases, capsules / fruits of with very small seeds that may dehise during transportation, close-weaved cotton bags should be used (Turnbull, 1975). Large open-mesh baskets are ideal for promoting free air circulation in cones and other large fruits and may be constructed from locally available materials, whether metal, willow, bamboo or rattan. Aeration in loose-piled fruits can be improved by inserting loosely-constructed "chimneys" of wooden slats in the middle of the piles. Daily turning of loose-piled fruits or of sacks can do much to improve access of air to the less exposed fruits.

If temporary storage is essential, then the seeds must be stored at 15-20°C in perforated poly bags for a short period provided they tolerate these temperatures.

If transport distances are short and large quantities of a single species and provenance are to be transported, fruits may be loaded into vehicles without containers (Morandini, 1962, Goor and Barney, 1976). The vehicle must be cleaned of all seed carried on previous trips before a new lot is loaded. Individual containers should be used for longer journeys or smaller seed lots. Sacks should be carefully arranged in the vehicle to allow as much air circulation as possible between them. Open-mesh baskets are excellent for promoting free air circulation during transport as well as during temporary storage.

Generally, for most species open trucks and trailers should be used in preference to closed vans in order to promote air circulation (Dobbs *et al.*, 1976). But for species which need to maintain high moisture content if they are to retain viability, care must be taken to prevent excessive drying; use of polythene bags and provision of shade against sunlight are necessary. Rapid transport directly to the final destination and immediately after collection may be essential for some species which either germinate naturally or lose their viability soon after seed fall at normal temperatures. Provision of special insulated containers, to control temperature and humidity in transit, may also be needed (Kemp, 1975). Styrofoam containers or metal thermos flasks are recommended as a precaution against temporary exposure of recalcitrant seeds to cold or freezing conditions in aircraft holds (IBPGR, 1981). This situation may be found among some trees of the ever-wet tropical high forest but is rare among species which have evolved under conditions which include a severe cold season or dry season each year. Many of the species most widely used in plantation forestry fall into the latter category.

It is important to ensure that the minimum possible time should elapse between the despatch of fruits or extracted seed from the collecting site or local seed collecting depot and its arrival at the central seed processing depot. Part of the planning operation must be to provide transport which is adequate in quantity and quality to avoid delays in despatch and breakdowns en route. Even short stop-overs add to the heat build-up in fruit and cone sacks during transit. Vehicle drivers should be informed of the nature of their loads and the need for proper care and prompt delivery (Dobbs *et al.*, 1976). Road transport is commonly used for at least the first part of the journey. Railway transport may be more economical over long distances (Morandini, 1962) and air transport quicker, but both involve some loss of control over storage conditions in transit and any trans-shipment means extra handling and delay. Road transport is likely to be the preferred method in most conditions.

With valuable collections it may be advisable to split each lot of seed or fruits into at least two parts, to travel separately, in order to guard against losing an entire collection through accidents en route. It may also be desirable to insure the contents against loss or damage, for a sum that will cover at least a proportion of the cost of repeating the collection (Kemp, 1975).

Information on the estimated time of arrival of a shipment of fruits must be sent in advance to the seed processing depot. This enables the receiving station to provide staff for prompt unloading of fruits (Dobbs *et al.*, 1976). Similar information should be sent to intermediate recipients responsible for trans-shipping or forwarding consignments of fruits.

3.6. Drying

Care must be taken prior to drying recalcitrant seeds. A temperature of 15°C is safe for most tropical species. Delay in drying or slow drying combined with high temperature is highly lethal to such seeds. Before subjecting the seeds to drying, they must be allowed to equilibrate by placing them at 15°C for 1 day if moist (above 20% MC) or for 2-3 days if dry (<20%).

The rate of drying is also dependent on seed quantity, depth of drying layer, circulation of air within the drying chamber and species. The seeds must be placed over as a thin layer on top of a metal mesh to maximize air circulation. Freshly dried or regenerated silica gel can be used to attain 5% RH at 20°C to dry the seeds to lower moisture contents (Hanson, 1985 and Cheng *et al.*, 1990). The first change of colour (reduced intensity of blue) occurs when silica gel is at equilibrium of 12-13% RH and when it is pale blue, it has 49% RH. One hundred gram of regenerated silicagel at 5% RH can absorb 7 g of moisture from seeds before the colour diminishes. The temperature is important hence the seeds along with silicagel must be placed in perforated boxes kept in incubators maintained at different temperatures. The silica gel must be regenerated 2-4 times a day when the seeds are fresh and moist and every day when the seed moisture content is reduced to 10-15% and every 3-7 days when moisture content is reduced below 7%. Silica gel can be regenerated by drying in an oven maintained at 130°C for 3-4 hrs. It must then be stored in sealed container overnight to cool at ambient temperature before used for seed drying.

In order to determine if the seeds have reached the safe moisture content, the following formula suggested by Hong and Ellis (1996) can be used:

$$\frac{(100\text{-initial MC\%})}{(100\text{-desired seed moisture content (DMC)\%}} \times \text{initial seed weight (g)}$$

This formula provides the weight of seeds, once they have reached the desired moisture level from the initial moisture level. A worked out example is provided for easy under standing.

Initial moisture content : 60%

Desired moisture content : 15%

Initial weight of seed sample : 150g

The weight of sample when dried

to the desired moisture content of 15% will be :

$$= \frac{100-60}{100-15} \times 150$$

$$= \frac{40}{85} \times 150$$

$$= \frac{6000}{85}$$

$$= 70.58\text{g}$$

Hence the seed sample has to be dried to 70.58g to help it achieve 15% moisture content that is the desired storage.

3.7. Determination of Seed Moisture Content

Procedures outlined by ISTA can be followed. In case of large seeds, the determination of moisture content of 20-25 individual seeds is recommended. Separate determination of moisture content of embryo, embryonic axes, cotyledons, endosperm and other seed covering structures will be necessary. Seed moisture is always expressed in wet / fresh weight basis and reported to nearest 0.1%.

3.8. Determination of Germination Percent

Germination test must be continued till the seeds germinate or rot. This duration can be as long as 18 months for some species. Removal of seed covering structures, filling, nicking or nicking with needle may promote germination. Tetrazolium test must not be used. However, as sparingly as possible it may be done to determine the effect of dormancy when seed remain ungerminable at the end of germination test. The Rules provided by ISTA is followed. However, the Hand book of Seed Technology for Gene Bank Vol. I (Ellis *et al.*, 1985) can also be used.

3.9. Humidification of Dry Seeds

Seeds maintained at 8-12% moisture can be subjected to imbibition damage if exposed directly to water. Hence, they must be slowly raised to seed moisture levels of 15-17% by placing them in humid atmosphere and done within 12-48 hrs at 20°C. Small seeds are spread uniformly over a petridish or over a box lined at the bottom with wet paper towels and closed. This entrapment is kept at 20°C. Large seeds were placed in netted bag over wire gauze above water in a desiccator and kept at 20°C. The seeds must be spread as thin layer for rapid absorption of moisture.

Additional precautions for collection of recalcitrant seeds in the humid tropics as follows (Bonner, 1996) :

Most of the problems in maintaining seed viability are accentuated in the case of recalcitrant species in the humid tropics. They have short life spans and can neither tolerate low temperatures (not much below 20°C) nor reduction in their moisture content below a relatively high value. The majority of seeds in the humid tropics are recalcitrant, and as a result of high rates of seed spoilage in transit, such species are seldom used in reforestation except within their countries of origin. Even for seeds collected and used locally, deterioration can be serious within a matter of days unless special precautions are taken. Collectors have to work within relatively narrow limits of tolerance.

3.9.1. Ventilation

Recalcitrant seeds (and their fruits) respire actively, hence require good ventilation. If large quantities are closely packed, suffocation, physiological breakdown, fungal growth and overheating will occur resulting in rapid death of the seeds. If plastic bags are used as containers, their tops should be left open or small holes should be made in their sides. Baskets or cloth bags are suitable although usually more bulky or expensive. It is not easy to strike the right balance between adequate ventilation and adequate moisture conservation because most of the seeds are highly sensitive to desiccation and ventilation though useful in dispersing metabolic heat, may be counter productive some times.

3.9.2. Temperature

Temperatures below 20°C or above 35°C should be avoided. Low temperatures are likely to be experienced in air transportation unless the seeds are kept in the pressurized cabin. High temperatures may be due to respiration or to direct heating by the sun. Good ventilation will reduce heat build-up from respiration. Recalcitrant seeds should be kept shaded from direct sun at all times.

3.9.3. Moisture content

Recalcitrant seeds deteriorate if their moisture content is reduced too much or too rapidly. This is likely to happen during transportation in open vehicles because of air movement. The size and number of ventilation holes in the containers should be reduced under such circumstances. Open containers should be covered with newsprint or cloth to reduce the desiccating effect of air movement.

3.9.4. Organization of nursery

Before the collection is made, the recipient nurseries should be forewarned to have their germination beds ready. Recalcitrant seeds should be sown as soon as possible after collection.

3.9.5. Long trips

Collection trips for recalcitrant seed should not exceed a few days in duration. If a long trip cannot be avoided, a lot of extra work has to be done in daily inspection and processing of the collections already made. If decay and fungal growth set in, the seeds will have to be spread out for better ventilation. Decaying pulpy fruits will have to be separated from the healthy fruits and depulped immediately. Capsules must be discarded as soon as they have opened sufficiently for their seeds to be extracted. If seeds begin to germinate during the trip, the germinants may be saved by storing in rigid containers or baskets lined with newsprint or other absorbent material and kept moist. Some seeds deteriorate so rapidly that the best way to transport them may be in a germinating condition in a moist medium.

3.10. Dormancy and Its Effects

Though several dormancy alleviation techniques have been provided, it is preferable not to pretreat seeds unless, it can be confirmed, that the method does not affect the seed. Seeds must be assessed for dormancy and knowledge of the type of dormancy (depends on the plant family). Seeds should not be scarified on any account either using acid or hot water. Scarification using sand paper or removal of covering structures by hand was preferred. While methods like prechilling, application of GA_3 (100 ppm), potassium nitrate (0.1%) and light can be used without damaging the seeds. Soaking seeds for 6hr in GA_3 50 ppm for clove and 100 ppm for nutmeg and rubber improved germination. $FeSO_4$ 1% was next best and KNO_3 0.5-1.0% was deleterious to seed germination (Gunasekaran, 1997). Neem seed treatment with Kinetin, GA_3, IAA at 200-400 ppm and fortification with 2% Potassium Nitrate improved seed germination. Bharathi (1999) observed that seeds soaked in 100 ppm GA_3 / 2% KNO_4 for 24 hrs or halogenated with chlorine based formulation at 4 g/kg for 7 days or osmoprimed in 0.25% Poly Ethylene Glycol (PEG) 6000 at 15°C and 1000 Lux Illumination for 1 hr or coated with *Albizia amara* dried leaf powder gave better germination and vigor. Jackfruit seed germination was higher at 25 - 30°C and improved when soaked in water for 24hrs. Washing seeds improve germinability. Treatment with 100 ppm GA_3 gave 100% and 77% in 25 ppm NAA compared to control. Soaking of phalsa (*Grewia asiatica*) seeds for 24 hrs improves germination (Srimathi and Sastri, 2004). Coffee cv. Tafaricala seeds were easiest to germinate in 55 days compared to 99-137 days for other cultivars and selections. Light exposure delayed germination by 5-7 days. Germination of fresh seeds was enhanced from 5% to 35% by slurry treatment with 1% 8 Hydroxyquinone.

In neem, germination potential did not vary among seed sizes, with large seeds showing lower rates of germination, but possessed higher values for seedling length, dry weight etc., Seeds sown with radicle side down had higher rate of germination, vigor and vigor index. The presence of parchment membrane delayed germination. Seeds soaked in 0.5% potassium di hydrogen orthophosphate for 16 hr and followed by air drying for 6hr was assessed after 90 days and revealed the positive effect of this treatment over control. Acid treated and decoated cocoa seeds germinate within 20 days compared to untreated seed (30 days). Seedling vigor was also higher (Ponnusamy *et al.*, 2007).

Hong and Ellis (2000) opined that dormancy (either primary or induced) was a positive character, because it helped in moist imbibed storage.

3.11. Effect of Fungal Infection

Fungal consequent to harvest or during maturation in an integral factor of a recalcitrant seed. The presence of sugars in fruit mesocarp, high seed moisture, hot and humid climate favor growth of microorganisms. The composition of microflora in and on the recalcitrant seed narrows down with storage (Mycock and Berjak, 1990). In arecanut drupe, species viz., *Fusarium spp., Penicillium spp., Aspergillus flavus,* A. *niger, Rhizopus stolonifer, Trichoderma spp.,* and *Botryodiplodia theobromae* were found associated (Raja, 2001). Mechanical injury caused during harvest, when fruits are dropped to ground, or during transport also increase the liability of fungal infection. The soil adhering to the fruits while collecting from the ground was a source of fungal infection (Nambiar and Koti Reddy, 1979). Fungal infection occurs during germination especially for those seeds with mucilaginous layer subjected to prolonged duration test and when seeds are dead. Biocides may be avoided and mucilaginous seeds are disinfected by soaking in 1% sodium hypochlorite solution for 2-5 min, dried and placed for germination. High initial seed moisture could play an adverse effect to seed treating fungicides, while lowering the seed moisture may be synergistic to fungicidal dressing. Ponnusamy *et al.* (2007) reported the adverse effect of thiram (2%) treatment for coffee cv. Tafaricala to combat *Fusarium* infection. They also reported that coffee beans at moisture contents of 20-40% lost viability within 3 months when treated with fungicide captan (2 g/kg). But when coffee beans were dried to 9%, seed treated with captan and stored in Paper Aluminum Foil (P-AF-P) pouches maintained viability at 20%, 45% and 68% for 6 months in storage at tropical, sub tropical and temperate regions. But only 19% and 54% viability was recorded in sub tropical and temperate regions while no seed survived in tropical conditions beyond 12 months. Fungicide application may also be deleterious in some cases. Thiram at 2% reduced germination in neem (Bharathi, 1999). However, internal infection by *Fusarium* spp. caused rapid loss of viability. Working with nutmeg, Madusudanan and Babu (1994) found that seeds treated with 0.1% mercuric chloride improved germination. Treating arecanut seed drupes with *Pseudomonas fluorescens* was found to reduce fungal contamination by *Fusarium* spp. (Raja, 2001) (Figure 3.3).

Contd..

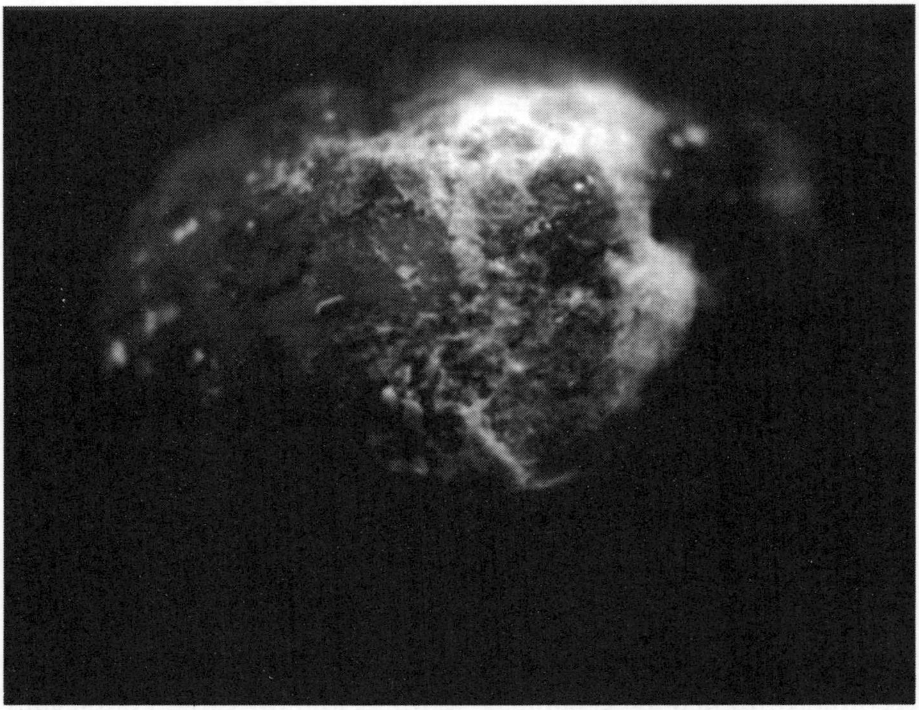

Figure 3.3 : Fungal infection of arecanut seed (Raja, 2001)

□□□

Chapter 4

Identification of Recalcitrant Seeds

There is a sudden urge to identify and preserve our natural resources considering the speed in which humans modify their environment. Vast tracts of prime forests are being felled to help humans to achieve their minimum standard of living. With this operation, several species are vanishing without being acknowledged for their presence. Search for new molecules by pharmaceutical companies have propelled them to areas beyond human reach and they are the ones now promoting conservation of natural resources.

One form of conservation is *in-situ* conservation, ie. leaving the habitats unscathed. But the ever growing demand for natural resources has put a huge pressure on conservation. The other way is *ex-situ* conservation, which demands preserving the gene in any manner possible as seed, tissue or DNA. According to Harrington (1972) of all the *ex-situ* conservation strategies, the easiest and least expensive method of preserving the world's existing plant genotype would appear to be conventional seed storage. It plays a complimentary role in germplasm conservation. Preserving the seed thus is the best method advocated by conservationists, as it is cheap and effective. Before going for conservation, the nature of seed must be known for forming the protocol.

Seed survival of more than 7000 species was collected and published as a *Compendium* by Hong *et al.* (1996), probably the first most exhaustive collection of seed storage behaviour for *ex-situ* conservation. But it was limited to only 2.5% of the 2,50,000 species of flowering plants. The Royal Botanical Garden, Kew has a vast data base of 10600 species whose storage behaviour has been catalogued. Of these, 540 species have been identified by Millennium Seed Bank as recalcitrant.

Lot of confusion prevails before labeling seed as recalcitrant or not. Barbedo and Bilia (1998) were concerned about the absolute nature of terming the seeds either recalcitrant or orthodox. But wanted to distinguish non-orthodox seeds based on the following features such as a) natural habitat, b) longevity during storage and its conditions, c) seed moisture content at 50% viability, d) chilling sensitivity, e) seed dormancy.

A thorough examination and classification of recalcitrant seeds through protocols that involves the determination on drying seeds to two or three different moisture levels and assessing per cent germination was listed out by Hong and Ellis, (1996) which will be dealt briefly in this chapter.

Some of the characteristics of recalcitrant seeds are provided below for easy assessment and categorization.

4.1. Seed Provenance

Upon receipt of the sample, examining the area of its collection will provide a clue to its nature. Generally species from tropical rain forest are desiccation sensitive. Savannah species may be desiccation tolerant to a varying degree. While palms from dryland / deserts show orthodox behavior. Even among drier habitats, species that shed their seeds during the wet season are more likely to be desiccation sensitive than those that shed during dry season. Again contrary to popular notion, most aquatic seeds produce orthodox seeds.

Roberts and King (1980) hypothesized that orthodox species originate from environments subjected to occasional drought where attainment of desiccation tolerance is a valuable asset in survival strategy. Recalcitrant species, originate in moist ecosystem in which seeds are subjected to high humidity during development, maturation and shedding. Thus among 115 shrub species originating from Mojave Desert show orthodox behavior (Kay *et al.*, 1988) while seeds belonging to *Dipterocarpaceae* that are native to dry areas are desiccation tolerant while those from tropical regions are sensitive. Thus *Arabica* coffee shows both characteristics and is intermediate as it was evolved in cool and dry region of Ethiopia, while *Liberica* coffee being a native of hot and humid region of Liberia in Africa is recalcitrant (Hong and Ellis, 1995). Dickie *et al.* (1992) proved that among *Palmae* family, species from the tropical region are recalcitrant while those from hot dry regions are orthodox. Though oak species is recalcitrant, *Quercus emoryi* which is a native of Savannah is not recalcitrant. An exhaustive list comprising of 7000 species and 251 families has been compiled by Hong *et al.* (1996).

In general, recalcitrant species can be sub-divided into those of tropical origin and those adapted to temperate climates and high altitudes of tropics. The latter can be stored at cooler temperatures for longer periods.

Hill and Edwards (2010) however found that dispersal periods of Australian recalcitrant species did not coincide with wet season which is the norm in tropical rain forests. Hence the recalcitrant behaviour of tropical rainforest tree seeds is chiefly due to they being shed during rainy season may not hold good for Australian recalcitrant species. They opined that onset of seasonal aridity on the other hand would lead to death of desiccation sensitive seeds.

4.2. Taxonomic Classification and Seed Storage

Species belonging to *Chenopodiaceae, Combretaceae, Astraceae, Labiatae, Solanaceae* and *Pinaceae* show orthodox behaviour. Species belonging to *Rhizophoraceae* are recalcitrant. Species within *Leguminosae, Graminae, Cucurbitaceae, Cruciferae* and *Rosaceae* are generally orthodox but have several exceptions. Several species of *Dipterocarpaceae* and *Sapotaceae* show both recalcitrant and intermediate storage behaviour. While species of *Meliaceae* show all three categories of storage behaviour.

Seed storage also varies among species level within a genus. Most species of genus *Acer* are orthodox but not for all species. Among *Auracaria* species, A. *angustifolia* and A. *araucana* from South America and A. *bidwillii* from Australia are recalcitrant. While A. *heterophylla* and A. *cunninghamii* and A. *columnaris* of the New Caledonian Island show orthodox behaviour (Tompsett, 1983).

Walters *et al.* (2008) opined that incidence of recalcitrance does not distribute among phylogenetic clades and some plant families include many species producing recalcitrant seeds viz., *Fabaceae, Lauraceae, Sapindaceae, Meliaceae* while some families lack species exhibiting recalcitrant behaviour viz., *Solanaceae, Asteraceae* and *Amaranthaceae*. Life history traits of the plant such as long lived, perennial nature, habitat such as aquatic or rainforest were associated with recalcitrance.

Further, species that produce drupes containing 1- 4 seeds or those that are arillate, pods containing 1-5 large seeds of many arillate seeds; berries containing 1-10 seeds; capsules containing 1-5 seeds and single seeded nuts show all three categories of storage behaviour.

It has been found that presence of aril on seeds is associated with short life span, does not affect storage behaviour. Similarly species showing deciduous nature do not show desiccation tolerance but do exhibit chilling tolerance. Rubber, classified as recalcitrant, can be stored at 7-10°C and other species of Dipterocarps viz., D. *alatus*, D. *intricatus*, D. *obtusifolius* and D. *tuberculatus* can tolerate low temperature of 6°C.

Leprince *et al.* (1992) observed that taxonomy did not have any influence on recalcitrance and suggested that it is a derived trait, with tolerance having been lost a number of times among the species within the taxonomy. Recalcitrant seeds were best represented in species from mesic tropics, especially among rainforest climax species, but it also appears among species in cooler, drier and seasonal environment.

4.3. Seed Morphology

Seed size *per se* does not determine seed storage behaviour, but generally speaking recalcitrant species are generally large. There is a strong association between thousand seed weight and seed storage behavior with a genus especially *Acer* and among several genera. However, the heaviest seeds of orthodox were recorded in *Hardwickia pinnata* (6300 g / 1000 seeds) and cashew (5000-8000 g /1000 seeds). It may be opined that species that possess thousand seed weight more than 13000 g show orthodox, intermediate or recalcitrant behaviour and thus at this range seed size does not provide any indication of seed storage behaviour. Again seeds weight less than 25 g also show orthodox behaviour even though exceptions within *Orchidaceae* and *Cruciferae* are found (Figure 4.1.).

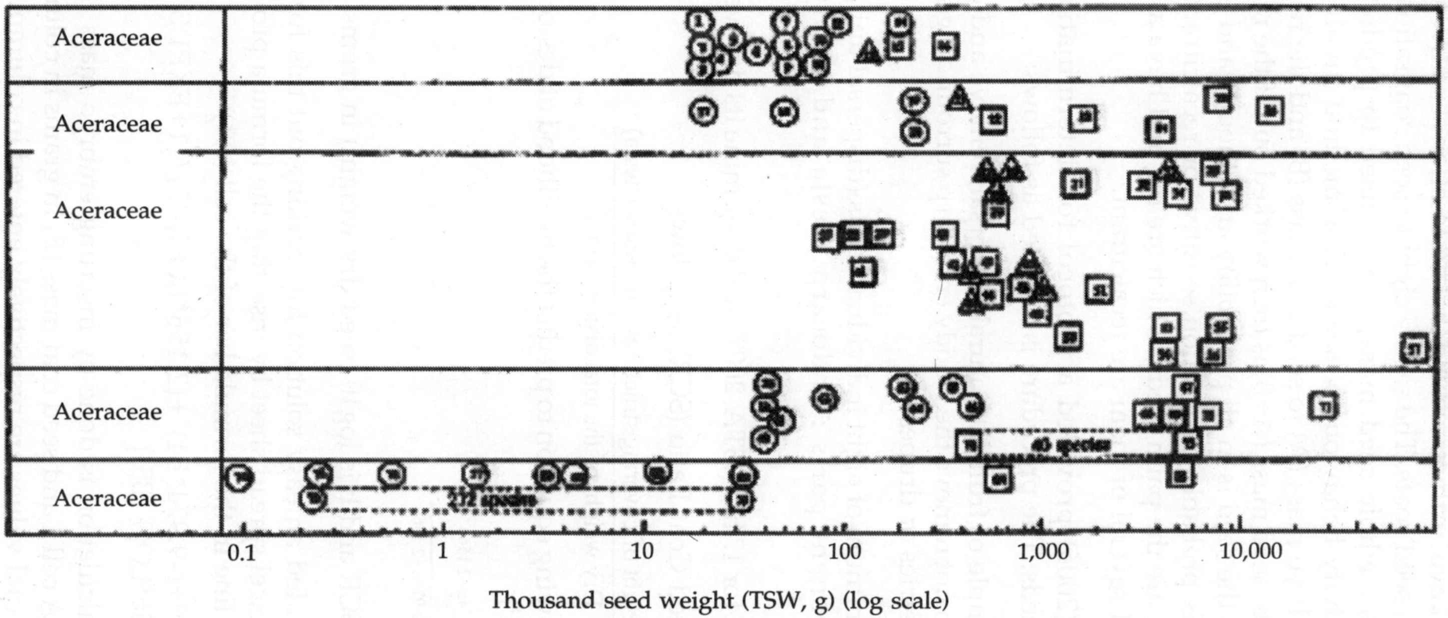

Figure 4.1 : Relationship between thousand seed weight (g) and association of recalcitrant seed behaviour in 52 species belonging to five families (circles show orthodox; triangles intermediate and squares recalcitrant storage behaviour).
Courtesy: Hong and Ellis (1996).

Observations have further revealed that desiccation sensitive seeds are large with thin seed coats. The seed weight to seed coat ratio (SCR = seed coat mass : whole seed mass) can be used to predict the desiccation sensitivity behaviour of seeds. This method involves in measuring the embryo mass (A) to seed coat mass (B) and the ratio of embryo mass (A) to seed mass (A+B) is then worked out. If the ratio is less than 0.5 then, the seed is in all probability an orthodox and if it is more than 0.5, it is probably desiccation sensitive or recalcitrant. An exception though, are the palm seeds, which are big and have a thick seed coat, but still several of them are recalcitrant.

Daws *et al.* (2006) provided a protocol for determination of recalcitrance in seeds. The procedure is outlined as follows:

a. Collect a sample of fruits at natural dispersal; remove and clean seeds, but do not remove the woody endocarp surrounding seeds formed in berries or drupes.

b. Dissect a minimum of eight individual seeds (dispersal unit) into their component parts : endocarp /testa and embryo / endosperm.

c. Dry at 103°C for 17 hrs (ISTA, 2007) and determine its dry weight.

d. Calculate Seed Coat Ratio (SCR) as follows:

$$\text{SCR} = \frac{\text{Dry weight of covering structures (endocarp + testa)}}{\text{Dry weight of the total dispersal unit}}$$

e. Use the following equation to predict the likelihood of desiccation sensitivity (P):

$$P = \frac{e^{3.269-9.974a+2.156b}}{1+e^{3.269-9.974a+2.156b}}$$

Where, a is SCR and b is log10 (seed dry weight) in grams.

He also provided an easy solution for working out this formula using Microsoft Excel spread sheet by inserting the formula provided below as a single line in the formula bar of the cell (A1):

= EXP [3.269+(-9.974*B4) +(2.156*LOG(B5)/(1+EXP) 3.269+(-9.974*B4) + 2.156*LOG (B5)]

The actual calculation is done by inserting embryo mass (A) in grams in column A cell 1 and seed coat mass (B) in grams in column A cell 2 while the actual values are respectively entered in column B cell

1,2. Enter the above formula while keeping the cursor in cell B7. The spread sheet will automatically determine the values in B4 and B5 and calculate the probability.

A conceptual view of the Excel spread sheet is provided below :

	A	B	C
1	A. Embryo mass (g) =	2	
2	B. Seed coat mass (g) =	2	
3			
4	Seed Coat Ratio (SCR)=	0.5 (B4)	A/(A+B)
5	Seed mass (g) =	4 (B5)	(A+B)
6			
7	Probability seed is recalcitrant (P) =	0.396526643	
8	If P<0.5 seed is probably desiccation tolerant / orthodox		
9	If P >0.5 the seed is probably desiccation sensitive / recalcitrant		

4.4. Seed Moisture Content at Maturity

Research has shown that seeds having moisture content of 36-90% at maturity are generally recalcitrant. While those having 23-55% are intermediate and those with moisture content below 20 and between 50% show orthodox behaviour. Hence, seeds that shed with moisture content above 60% are generally recalcitrant (Figure 4.2).

Relationship between seed moisture and thousand seed weight reveal that even if the thousand seed weight is high, lower shedding moisture content will generally result in orthodox behaviour while seed moisture content between 23-70% and thousand seed weight of 1000-7000g will show intermediate and those above 70% with a thousand seed weight of more than 1000g are typically recalcitrant (Figure 4.3).

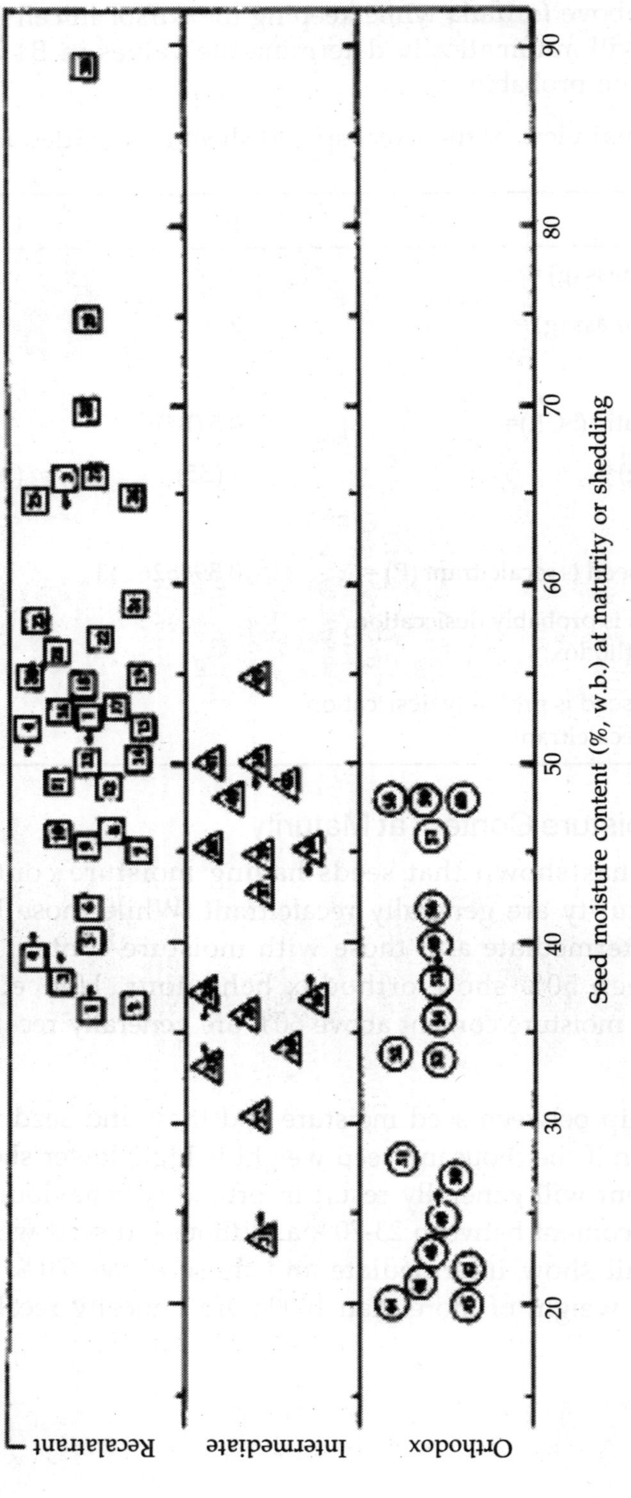

Figure 4.2 : Relationship between seed moisture content at maturity for 60 species and seed storage behaviour. Circles show orthodox and triangles intermediate and squares recalcitrant behaviour.
Courtesy: Hong and Ellis (1996).

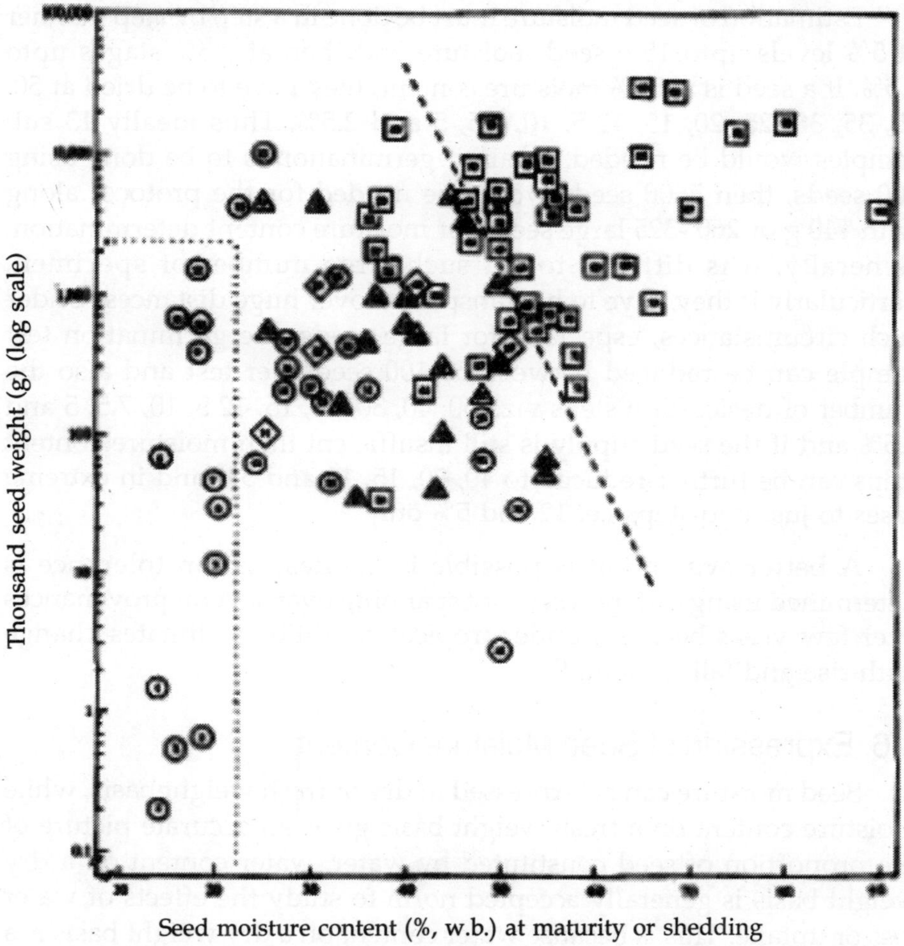

Figure 4.3 : Relationship between seed moisture content and thousand seed weight. (Circles show orthodox and triangles intermediate and squares recalcitrant behaviour). *Courtesy: Hong and Ellis* (1996).

4.5. Identification of Seeds Using their Ability to Withstand Desiccation

Long term storage means, a seed can be successfully stored at -20°C with 5±1% moisture content, medium term storage at +20°C and 40-50% RH and 7-11% seed moisture content or short term with seeds imbibed at +15°C.

Determination of seed recalcitrant behaviour is subject to error due to the degree of seed maturity. Hence the samples must first be identified by their maturity stage and then proceeded for drying.

Estimation of seed moisture must be done in a step by step manner at 5% levels upto 15% seed moisture and then at 2.5% stages upto 2.5%. If a seed is at 58% moisture content, they have to be dried at 50, 45, 35, 30, 25, 20, 15, 12.5, 10, 7.5, 5 and 2.5%. Thus ideally 13 sub samples would be needed. Again if germination is to be done using 400 seeds, then 5600 seeds would be needed for the protocol along with 140 g or 260 -325 large seeds for moisture content determination. Generally, it is difficult to get such large number of specimen, particularly if they have to be transported over huge distances. Under such circumstances, especially for large seeds, the germination test sample can be reduced between 50-100 seeds per test and also the number of desiccation steps viz., 50, 40, 30, 20, 15, 12.5, 10, 7.5, 5 and 2.5% and if the seed supply is still insufficient then moisture content steps can be further reduced to 40, 20, 15, 10 and 5% and in extreme cases to just two steps i.e. 12 and 5% only.

A better evaluation is possible if the desiccation tolerance is determined using different seed lot spanning over several provenances over few years because, under tropical conditions, climates change with rise and fall in rainfall.

4.6. Expression of Seed Moisture Content

Seed moisture can be expressed in dry or fresh weight basis, while moisture content on a fresh weight basis gives an accurate picture of the proportion of seed constituted by water, water content on a dry weight basis is generally accepted norm to study the effects of water loss or uptake. This is because water content on a dry weight basis is a linear expression of the quantum of water in plant tissues.

As mentioned earlier, seed moisture on dry weight basis is a linear relationship. A seed with moisture of 0.2 g water/g dry matter is twice as hydrated as that with a seed moisture estimated as 0.1g water/g and four times as mush as a seed at 0.05 g water/g dry weight. Hence it is a linear relationship.

Seed moisture estimated at wet weight basis is a non linear relationship, because, during the moisture estimation, the formula has the seed fresh weight both as a numerator (fresh weight-dry weight) and as a denominator (fresh weight). Thus, the moisture content is the reciprocal of fresh weight. Hence, when the seed moisture with 80% is dried to 70% and 60% on fresh weight basis, the seed actually looses 41.7% and 62.5% and not 12.5% and 62.5% of the initial moisture. In fact the moisture lost from a seed dried from 80% to 70% is twice as much lower when it is reduced from 70 to 60%. This is because the loss of seed moisture depends on four factors (Pammenter *et al.*, 2002 and Sun, 2002) viz., a) the gradient in water potential between tissue and external air (i.e. Seed moisture vapour pressure and relative humidity of the atmosphere) ; b) the thin layer of saturated air surrounding the seed, which depends on the size and shape of the seed and the rate of air movement around the seed; c) the surface area: volume ratio, which is influenced by the size and shape of the tissue; and d) the hydraulic conductivity of the seed, which is affected by the chemical composition and the presence and nature of physical barriers like testa/pericarp, comprising the outer layer of the seed. Hence, the response of a seed to moisture loss depends on its drying rate and quantification of this drying rate is thus important. The drying rate is not linear, which rapid drying during the early phases and slowing down subsequently.

Liang and Sun (2000) and Sun (2002) found that while drying cocoa embryonic axes, the rate of water loss could be quantified as the rate constant of the exponential relationship and reported a decline in relative water content of embryonic axes excised from recalcitrant seeds was an exponential function and the inverse exponential function RWC = b/(β + r) to be more appropriate for the drying rate envisaged (where RWC is water content relative to initial value, β is a constant and r is the drying time, b is the time taken for the water content to be reduced to half its initial value and thus a minimum b value indicates rapid drying.

Tobias and Pammenter (2006) experimenting the drying rates of embryonic axes excised from six different species, opined that, under slow drying conditions, the relationship between drying rate has a high degree of correlation. However, embryonic axes of *Trichilia dregeana* (*Meliaceae*) offered transfer resistance to water indicating the role of embryonic tissue in drying rather than the difference between water vapour pressure and surrounding air. Seed size also significantly affects the drying rate, when size differences are smaller, the effects are less effective and the seed coat or cover effect becomes more pronounced. Hence, the constant of the modified inverse function appears to be better than the rate of constant of an inverse function, particularly under rapid dehydration conditions (Table 4.1).

Table 4.1 : β values and regression coefficient r^2 values for embryonic axes of six species when dried either slowly or rapidly.

Species	Rapid drying				Slow drying			
	β	r^2	b	r^2	β	r^2	b	r^2
P. sativum	-0.115	0.80	1.20	0.99	-0.017	0.71	8.33	0.73
Q. robur	-0.130	0.93	1.15	0.98	-0.006	0.94	90.91	0.98
S.madagascariensis	-0.308	0.61	0.17	0.76	-0.041	0.68	2.61	0.74
T. emetica	-0.861	0.80	0.11	0.97	-0.033	0.70	4.81	0.85
T. dregeana	-0.611	0.97	0.29	0.94	-0.016	0.94	11.11	0.99
A. marina	-0.034	0.99	19.16	0.99	-0.006	0.95	50.00	0.86
		0.85		0.94		0.82		

4.7. Determination of Desiccation Tolerance Using Germination Test

Upon drying the samples, the viability of seeds must be determined using germination test. It is essential to avoid imbibition injury to dry seeds and thus prevent the confusion that will arise out of dormancy. The germination test will provide any of the following results :

a. all seeds tolerate desiccation upto 5% moisture content and below and in particular to levels of moisture content in equilibrium with 10-12% RH at 20°C thus showing orthodox behaviour.

b. most or all seeds tolerate desiccation to about 12.5% seed moisture and are in equilibrium with 40+50% RH at 20°C and further desiccation to lower moisture contents reduces viability and thus possess intermediate storage behaviour.

c. most or all seeds are killed by desiccation to 15-20% moisture content are in equilibrium with >70% RH at 20°C in thus may possess recalcitrant behaviour.

The procedure provided above are preliminary and there may be some deviations arising out of maturity stages, extraction procedures like fermentation, moist storage and preparation of seeds for transport such as pretreatment and transport at low temperatures.

Drying rates also play a crucial role in determination of recalcitrance. Under slow drying conditions, plant tissues stay longer at intermediate water contents, permitting damage associated with disturbed metabolism to accumulate. But, with fast drying, tissues pass through the intermediate water content range rapidly, reducing damage accumulation. Thus excised embryos or embryonic axes of recalcitrant seeds can survive to lower water contents when dried rapid than when dried slowly (Pammenter and Berjak, 1999).

Combining all of the above, Hong and Ellis (1996) developed a protocol to determine seed storage behaviour the flow chart is provided below (Figure 4.4).

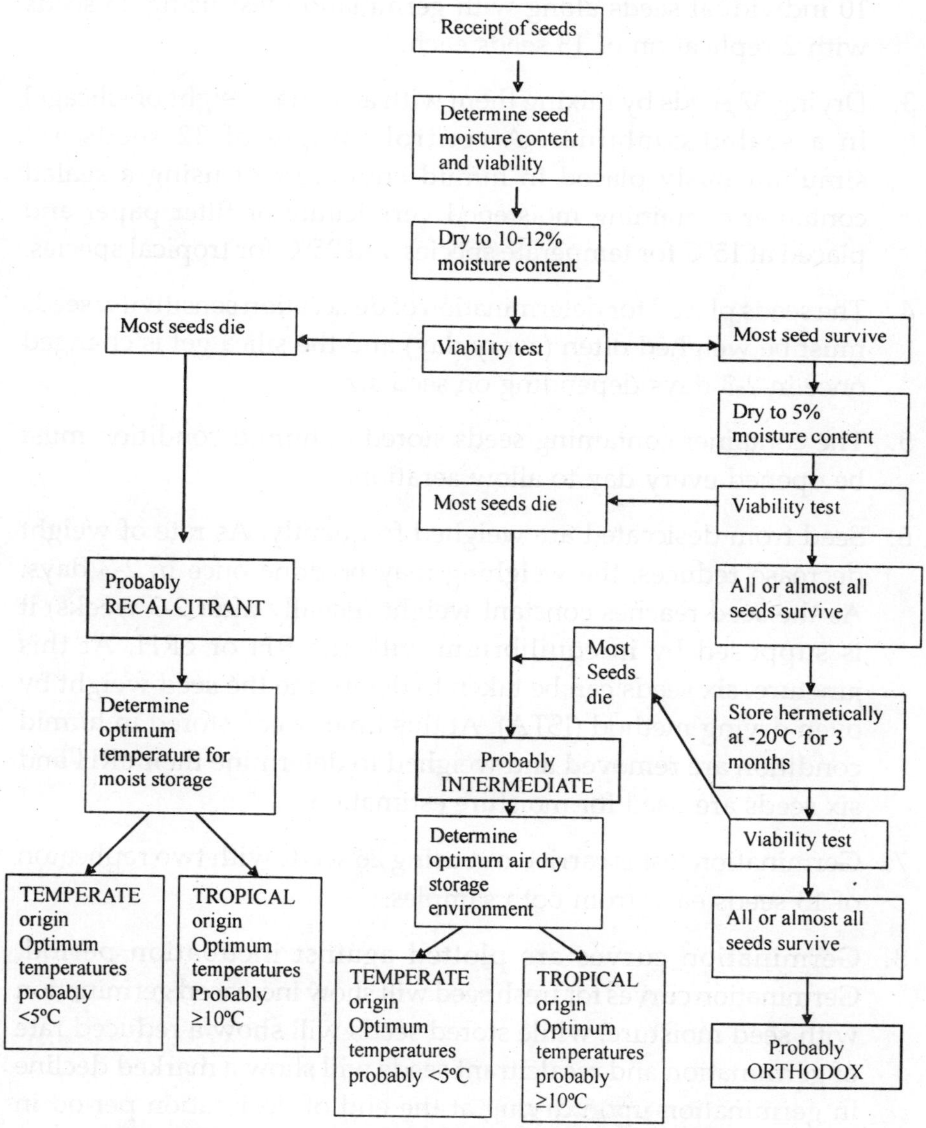

Figure 4.4 : Protocol for determination of recalcitrant behaviour.

Pritchard *et al.* (2004) provided a simple and elegant method (Fig. 4.5) to determine the recalcitrant behaviour of seeds as follows:

1. Determination of initial equilibrium relative humidity (eRH) of the seed lot with a hygrometer, using sufficient seeds filled in a sealed chamber.

2. Determination of initial seed moisture as per ISTA standards of 10 individual seeds along with germination test using 26 seeds with 2 replication of 13 seeds each.

3. Drying 32 seeds by mixing them with an equal weight of silicagel in a sealed container. A control sample of 32 seeds are simultaneously placed in humid environment using a sealed container containing moistened vermiculite or filter paper and placed at 15°C for temperate species and 25°C for tropical species.

4. The seeds placed for determination of desiccation sensitivity, seeds must be weighed often (every day) and the silica gel is changed once in 2-3 days depending on seed size.

5. The container containing seeds stored in humid condition must be opened every day to allow aeration.

6. Seed from desiccated are weighed frequently. As rate of weight decrease reduces, the weighing may be done once in 2-3 days. As the seed reaches constant weight (usually after 2-3 weeks) it is supposed by in equilibrium with the RH or eRH. At this juncture, six seeds can be taken to determine the seed weight by oven drying method (ISTA). At this time, seeds stored in humid condition are removed and weighed to determine their eRH and six seeds are used for moisture estimation.

7. Germination test is carried out using 26 seeds with two replicaton of 13 seeds each from both samples.

8. Germination curves are plotted against incubation period. Germination curves for fresh seed will show increased germination with seed moisture, while stored seeds will show a reduced rate of germination and recalcitrant seeds will show a marked decline in germination upon drying at the end of desiccation period in compared to seeds stored in humid condition (Figure 4.5).

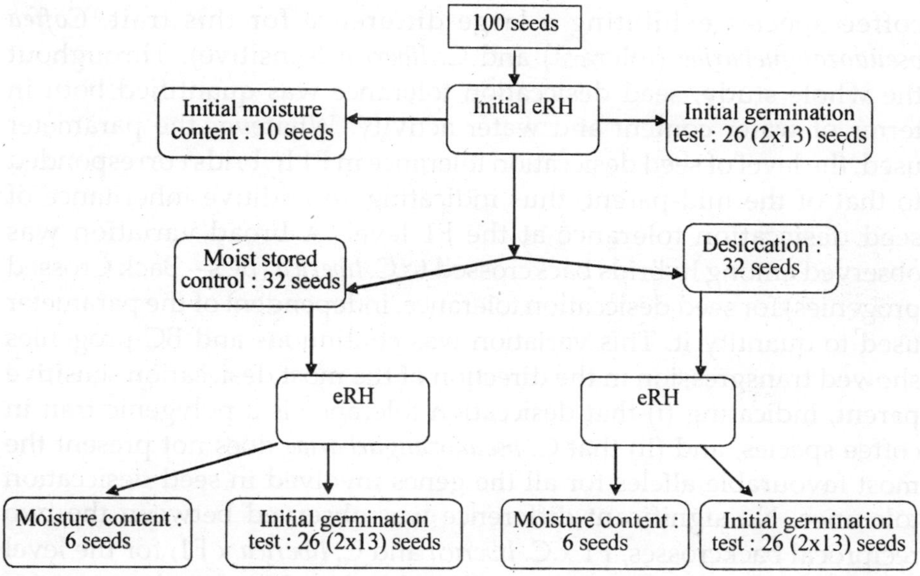

Figure 4.5 : Schematic representation of determination of recalcitrant seed (Pritchard *et al.*, 2004)

Pritchard *et al.* (2004) cautioned against false negative result which may result from desciccation tolerance screening when seeds that do not germinate after desiccation should not be confirmed as desiccation sensitive as the failure to germinate may be due to immature seeds, which are generally desiccation sensitive and several orthodox seeds are desiccation sensitive during their immaturation stage. Hence only mature seeds must be used for determination of recalcitrance. This is possible only when the biology of seed is known. Another cause may the role of dormancy. Use of alternating temperatrures, with light and dark sequences must be provided. If the seeds remain ungerminated even after this, a cut test may be performed to determine the viability of embryo or tetrazolium test may be performed. This has to be followed by use of appriate dormancy breaking methods.

Seeds of moist humid tropics may be viviparous and thus are highly sensitive to desiccation and may not germinate at all. Such seeds must be subjected to above test immediately upon receipt of seeds.

4.8. Genetic Association of Recalcitrance

Is there a genetic link to recalcitrant behaviour? Studies in these aspects are very few. Dussert *et al.* (2004) worked on the genetic determinism of seed desiccation sensitivity using a cross between two coffee species exhibiting a large difference for this trait, *Coffea pseudozanguebariae* (tolerant) and C. *liberica* (sensitive). Throughout the whole study, seed desiccation tolerance was quantified both in terms of water content and water activity. Whatever the parameter used, the level of seed desiccation tolerance in F1 hybrids corresponded to that of the mid-parent, thus indicating an additive inheritance of seed desiccation tolerance at the F1 level. A broad variation was observed among hybrids backcrossed to C. *liberica* (BCs – Back Crossed progenies) for seed desiccation tolerance, independent of the parameter used to quantify it. This variation was continuous and BC progenies showed transgression in the direction of the most desiccation sensitive parent, indicating (i) that desiccation tolerance is a polygenic trait in coffee species, and (ii) that C. *pseudozanguebariae* does not present the most favourable alleles for all the genes involved in seed desiccation tolerance. No significant difference was observed between the two reciprocal backcrosses, F1 x C. *liberica* and C. *liberica* x F1, for the level of desiccation tolerance of their seeds, showing the absence of a maternal effect on this trait. There was no significant effect of the number of seeds harvested from each BC on the level of desiccation tolerance of its seeds. Moreover, there was no significant correlation within BCs between seed size, seed viability, and water content before desiccation and desiccation tolerance.

❑❑❑

Chapter 5

Protocols for Safe Storage of Recalcitrant Seeds

Collection and classification of recalcitrant seeds is only the first step in *ex-situ* conservation. Saving the seeds for future, forms the most important and crucial aspect of conservation. Methods to determine the safe storage technique were outlined by Ellis and Hong (2006).

5.1. Determination of Seed Storage Environment

Seeds that can tolerate desiccation upto 5% does not mean they are orthodox. Because, such extreme desiccation may some times harm the seed also. Seeds of *Cattleya aurantiaca* (*Orchidaceae*) tolerated desiccation to 3.7 and 2.2% moisture content and germination was 94%. After 90% of sealed storage, the germination was 10% and only 36% after 6 years at 5ºC. Thus a factorial combination of three levels of moisture content (10, 7.5 and 5%), four temperature regimes (10, 5, 0, -20ºC) and two storage levels (3, 12 months of storage) under sealed conditions are needed for which 27 sub samples must be used. If sub sample are large enough, the storage duration can be extended. The sub samples are small then only one storage environment i.e. -20ºC and 5% seed moisture can be used. In case, the seeds survive the pre-storage desiccation treatment and do not survive during the subsequent storage period of one year, then it is mostly shows intermediate behaviour and if no loss of germination is noticed during the storage period, then it is mostly orthodox.

5.2. Determination of Optimum Air Dry Storage Environment of Intermediate Species

This is done by using a total of 105 treatment combinations at five different moisture levels viz., x+4, x+2, x, x-2 and x-4 where x is seed moisture content estimated during initial desiccation tolerance investigation. Similarly seven storage temperatures viz., 20,15,10,5,0,-10 and -20°C and three storage periods viz., 3,12 and 24 months should be tried to determine the best seed moisture content and best combination of temperature and storage period and also the most probable storage duration beyond which regeneration is required. A pattern of seed germination will emerge from this storage studies as shown in the figure provided below. Where a is a seed with orthodox nature while c is intermediate and e showing recalcitrant storage behaviour. Additional pattern viz., b and d may emerge if the seed was harvested prematurely or very late or has been subjected to unfavourable seed treatmen (Figure 5.1).

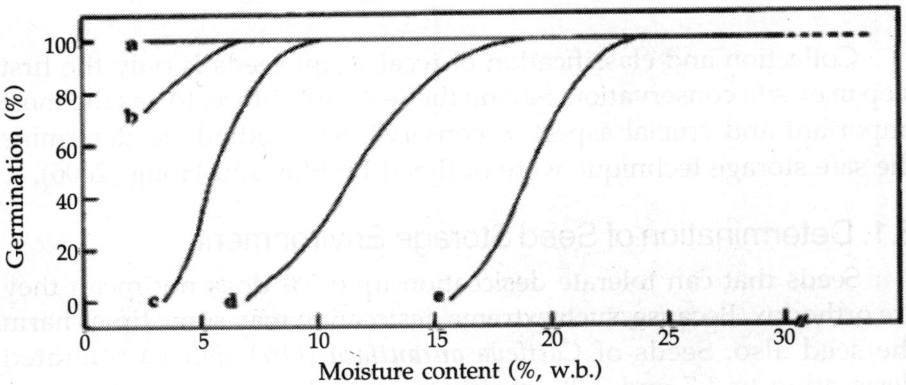

Figure 5.1 : Germination as affected by desiccation.
Courtesy: Hong and Ellis (1996).

5.3. Determination of Safe Storage Period for Recalcitrant Seeds

5.3.1. Viability estimation

Freshly harvested seeds are tested for germination at constant temperatures ranging from -5 to 30°C for species from temperate climates and from 5-40°C for species extracted from warm tropics at

5°C increments viz., -5, 0, +5, 10, 15, 20, 25, 30, 35 and 40°C for a minimum of three months to allow the seeds to germinate in cool temperatures. At the conclusion of germination test, the total number of rotten, firm and fresh ungerminated seeds has to be ascertained.

5.3.2. Chilling sensitivity

At the end of germination test, the firm, fresh ungerminated from each temperature regime must be transferred to optimum temperature for assessing the germination and the progress of germination. The optimum temperature is estimated by the first step in determining the nature of seed. This test provides the coolest temperature for storing without chilling injury or germination or death of the stored seed. To reduce the damage caused by inborne fungi, the seeds can be stored in temperatures less then that at which it is shed or fungicides or antibiotics can be used. A combination of low moisture levels ie., fully imbibed and subimbibed (2.5% less moisture than fully imbibed) and four temperature regimes viz., 20, 15, 10 and 5°C for species from low land tropics and 15, 10, 5 and 0°C for those from highland tropics and finally 10, 5, 0 and -5°C for species adapted to temperate latitudes for minimum of four storage durations viz., 1, 3, 6 and 12 months with a possible extension of 24 months for seeds adapted to temperate climate are recommended for assessing the storage of chilling sensitive seeds. Moist seed storage, at their shedding moisture content, in a moist medium is preferable to imbibed storage in aerated polythene bags.

5.4. Factors that Affect Interpretation of Seed Storage Behaviour

5.4.1. Use of immature seeds

Immature seeds with high moisture content (60%) are sensitive to damage during desiccation though their mature seeds are orthodox. At later stages of seed development such as at the end of seed filling phase, many orthodox species will show damage upon desiccation to low moisture levels. Hence, tests on immature seeds will result in erroneous interpretation of seed storage behaviour. Such seeds are common among tree species, where ripe seeds are lost by wind, insects, animals and pests and in species where hard seeds are found and when premature harvest is the rule. Delaying harvest especially in hot environment leads to higher presence of hard seeds.

5.4.2. Seed extraction methods

Seed extraction methods such as soaking, fermentation of fleshy fruits and storage in moist conditions especially during transport that involve long duration exposure to high moisture contents leading to the initiation of germination process and thus lead to poor storability.

5.4.3. Pretreatment

Seeds of temperate areas possess strong dormancy and prechilled ie. are stored under moist and cool conditions, to break dormancy. Prechilling of such seeds may lead to loss of desiccation tolerance due to initiation of germination process.

5.5. Drying Methods

Drying methods has to be done carefully especially for those seeds that possess high seed moisture at harvest leading to loss of viability earlier than necessary.

5.5.1. Desiccation

Some seeds like Arabica coffee, papaya and oil palm though show intermediate seed storage, are able to tolerate desiccation upto 10-12% and there may be several others whose desiccation behaviour are yet to be determined. Doijode (1990) while desiccating seeds of mango cultivars *viz.*, Alphonso and Totapuri, observed that seeds desiccated using silicagel did not germinate while those desiccated with fused Calcium Chloride gave 67% germination after 30 days storage. Hence, drying / desiccation methods must be perfected before seeds are taken for storage.

5.5.2. Delayed germination tests

Several of the recalcitrant species show either slow germination or dormancy, thus assessing the germination, a cumbersome process. Seeds of *Persoonia comata* need one year to germinate (Fox *et al.*, 1987), *Acrocomia sclerocarpa* (2.5 years), *Corylus avellana* (Jarvis, 1975). *Citrus* sp. (King and Roberts, 1980) are difficult to germinate when fresh. Several of the weed and shrub species need warm stratification, prechilling for germination (Young and Young, 1992). Seeds of *Anisophyllea griffithii, Hydnocarpus woodii* and *Barringtonia macrostachya* require 95, 83 and 65 weeks to attain 50% germination (Soepadmo, 1989). Other tests like immersion in LN for quick assessment of recalcitrant behaviour are also faulty as seeds of *Juglans, Prunus* and *Populus* that show orthodox behaviour are irreversibly damaged when placed at -40°C.

□□□

Chapter 6

Storage of Recalcitrant Seeds

The longevity of recalcitrant seed is very short; it varies from a few days to a few months or a year under proper storage conditions (Chin, 1989). Therefore recalcitrant seeds are to be sown immediately after collection to provide maximum germination. In general, most of the above said horticultural crop seeds are sown immediately after collection. Because of their short longevity, which varies from few days to few months, for example the arecanut seed looses its viability within 24 days under ambient (open) condition (Raja *et al.*, 2002 a) and 17 days for jackfruit (Shylla Merlin and Palanisamy, 2000) they may be kept within the fruit itself for few days. But due to the high moisture content it may enhance the pathogen entry and create germination loss. Preserving cocoa seeds in pods at ambient temperature of 27ºC was not found suitable as seeds lost viability completely within ten days due to fungal infection, high sugar content of pulp and presence of germination inhibitor (Sudhakara *et al.*, 2000).

There can be no doubt in the mind of any investigator working on desiccation - sensitive (recalcitrant) seeds that microorganisms, more particularly fungi, play a significant role in post harvest deterioration

(Berjak, 1995). Fungal species viz., *Fusarium* spp., *Penicillium* spp., *Aspergillus flavus*, A. *niger*, *Rhizopus stolonifer*, *Trichoderma* spp. and *Botryodiplodia theobromae* were found associated with the arecanut seed (Raja *et al.*, 2002 b). The relatively high moisture content, sweet nature of pulp and temperatures that tropical recalcitrant behaviour demands also favour profuse growth of pathogenic (and nonpathogenic) microorganisms (Bonner, 1995). The composition of the microflora on and in recalcitrant seeds has been found to narrow with increasing storage period (Mycock and Berjak, 1990). Treatment using fungicide (2g /kg carbendazim) had some benefits for storing jackfruit seeds for 56 days at 10°C (Shylla Merlin and Palanisamy, 2000).

6.1. Storage Methods

Recalcitrant seeds are named as such because of their difficulties in handling and storage. These seeds do not tolerate desiccation and low temperatures, as detailed earlier, both pre requisites for optimal storage. Maintaining high seed moisture content and storing seeds in containers that allow some gas exchange is important to preserve the viability. This method, involving the storage of seeds in various gases or in sealed containers or by waxing, has had some success (Chin, 1989).

6.2. Effect of Containers

Almost all the researchers had the same opinion regarding storing seed in tin containers *per se* which resulted in poor storage due to improper ventilation, accumulation of carbon-di-oxide and increased metabolic heat. Canvas bag also had no takers. Storage in poly bags was considered more favorable if seeds were placed in a moist media and kept in low temperature (Shylla Merlin and Palanisamy, 2000 and Srimathi *et al.*, 2001). As poly bags of thickness between 0.1 to 0.25 mm allow diffusion of gases but not moisture, they are the preferred storage containers of most researchers (Bonner, 1996). Polybags were reported to store seeds effectively if the ambient temperature was reduced. Nutmeg seeds can be stored in polybag for 45 days at ambient temperature. Thus Gunasekaran (1997) reported that clove, nutmeg and rubber seeds could be stored for 45, 70 and 85 days in polybag at 10°C. Neem stones and kernels stored in polybag (300 guage) at 10°C maintained a mean germination per cent of more than 60% for 6 months, 4 months at 5°C and 3 months at ambient conditions at a safe moisture content of 10-15% (Selvakumar, 1999). Neem Seeds could be dried to 10% and stored at 5°C but lost viability at -5°C on 120th day due to chilling injury. Neem seeds dried to 9% moisture content could be stored in cloth bag in ambient conditions for 2 months. Seeds needed for longer storage can be transferred to mud pots and buried in moist coir pith bed retained seeds for 6 months (Ponnusamy, 2003). Storage as stone was better than kernel irrespective of temperature or moisture at 10% packed in polybags (700guage) at 10°C for 6 months (Bharathi, 1999). Seeds of mango cv. Desheri gave 50% germination after a year when stored in 100 guage polybags after treatment with 1% 8-hydroxyquinoline sulphate and stored at 15°C (Srimathi and Sastri, 2004).

Roy and Khurdiya (1986) suggested construction of a zero energy cool chamber for storing arecanut seeds. This chamber was constructed under shade using a double walled brick with the interspace between the walls filled with sand. The floor was made of bricks and space between bricks filled with sand. The whole contraption was drenched with water until saturation. A bamboo lattice was used to cover the top and lined with gunny which was also sprinkled with water. Water was sprinkled every morning and evening to maintain the temperature at 23°C and RH 90%.

Storing fresh seeds of arecanut mixed with 5% moist sand or 0.2 M potassium dihydrogen phosphate premixed with sand at 5% level (seed: sand ratio 1:3) and packing loosely in 350 gauge polythene bags and stored in zero energy cool chamber recorded higher germination (85%) after four months of storage (Raja, 2001). The main reason attributed for the success of zero energy chamber was the reduction of temperature by 20°C within the chamber in comparison to outside and retention of seed moisture.

6.3. Effective Methods of Storing Recalcitrant Seeds

Recalcitrant seeds can be stored by the following methods:

a) Moist or imbibed storage in sand, sawdust or peat

b) Partial dehydration

c) Controlled atmospheric storage

d) Cryopreservation and *in vitro* conservation

e) *Ex-situ* conservation

Among the four methods listed above, the first three methods are short-term storage methods. Now a days cyropreservation and *in vitro* conservation has been gaining momentum in respect of long-term conservation of recalcitrant seeds.

6.3.1. Moist or imbibed storage

At present, successful short-term storage method is limited to moist storage (Chin, 1989). Moist storage has been practiced for many years for a number of crops, including rubber (Ang, 1977), rambutan in sterilized wool (Chin, 1975), mango in saw dust (Parisof, 1988), cocoa (Evans, 1953) in moist sand and moist sphagnum moss for arecanut (Nagwekar *et al.*, 1997). To avoid the desiccation sensitivity, the recalcitrant seeds are to be stored in moist media like damp charcoal, saw dust, moist sand and other moisture conserving materials and chemicals. This is a short-term storage method and the viability can be maintained up to three months or little bit more. Many workers had the successful storage with moist incubation treatments, eg. Bernard (1950) in *Shorea* spp., Chacko and Singh (1971) in mango, Ang (1977) in rubber, Patil *et al.* (1986) in mango, Gunasekaran and Krishnasamy (1998) in rubber and Szezotka (1975) for oak at -3°C for three years.

Moist storage of jackfruit seeds in screw cap bottle kept at 16-18°C was found to be ideal for preservation upto 45 days (Fernandez, 1982). Mixing fresh seeds of clove, nutmeg and rubber with sand or saw dust containing Jalsakthi (a propriety petroleum product that has high water holding capacity) 10% and retaining them loosely bound in polythene bags maintained viability of 25-30% for 3 months, 30-31% for four months and 38-39% for four months in clove, nutmeg and rubber respectively (Gunasekaran,1997). Mixing neem stones and kernels with sand or charcoal containing Jalsakthi 10% and placing them in loosely tied 300 guage polybags maintained the seed viability for 6 months. Sand was better than charcoal (Selvakumar, 1999). Viability of jackfruit seeds can be maintained upto 60 days at 25% when stored in charcoal in 1:1 ratio and stored at 5°C.

Fresh seeds of cocoa, mixed with 5% moist sand and impregnated with Jalsakthi 10% and stored in 700 guage polybag maintained viability during for 40 days or charcoal at 10°C but for 32 days at ambient temperature (Vanitha *et al.*, 2005). In Jamun, Srimathi (1997) reported extended storage for 3 months using 2% moist sand and kept at 10°C. Seeds of mango remained viable for 4-5 weeks after extraction. Storage in charcoal in a dessicator at 50% RH gave 80% germination after 70 days. It was better than storing in polythene bag, refrigerator, desiccator or in open jar. All seeds were dead at 3°C. Seeds retained 17.5% germination after 120 days when kept at 20-23°C with charcoal in sealed bags. Alphonso stones remained viable at 40% upto 90 days at 25°C and was completely dead at 8°C. Application of abscisic acid at 10^4 M did not prevent sprouting and treatments like cryoprotectants like DSMO 10% or glycerol did not help in overcoming chilling injury (Srimathi and Sastri, 2004). Studying the phenomenon of spongy tissue in Alphonsa mango, Ravindra *et al.* (2010) reported increased germination activity of mango seed that led to desiccation of moisture from the surrounding mesocarp leading to sponginess. Application of Placobutrazol (2.5%) during fruit maturity leads to reduction in sponginess. Placobutrazol was found to counteract the effect of high Gibberellic acid found in mango stone, which is the cause for germination initiation. This method can be advocated for mango seed collection and storage.

The moist storage treatments are found very effective in counteracting physiological deterioration of seeds (Bhattacharyya and Basu, 1992). In imbibed storage, the seeds are stored under water but only for a short period, for example the rubber seeds stored under water recorded only 60% germination, after one month of storage (Ong *et al.*, 1983). Viability of loquat seeds (*Eribotrya japonica* Lindl.) declined rapidly when stored at 5°C but when done so in moist conditions at high RH gave 92% germination after six months (Zink and Ojima, 1965). Among various moistened media tried viz., 5% or 10% saw dust, 5 or 10% coir pith, 10% activated charcoal, 10% peat and 2, 5 and 10% sand, seeds of jamun (*Syzygium cuminii*) were successfully stored for three months (76-84%) in 2% moist sand (4:1 sand : seed ratio) packed in 300 guage polybags (Srimathi *et al.*, 2001), while seeds lasted just a month without media. But in this method the problem is the fungal growth during storage. Therefore it is necessary to give some chemical treatment to control the pathogens during storage.

Also the seeds incubated with moist media may sometimes lead to *in-situ* germination. Nevertheless, these sprouted seeds can also be used for sowing because it will produce good seedlings as had been reported in *Synphonia globulifera* (Corbineau and Come, 1988), *Artocarpus heterophyllus* (Bhattacharyya and Basu, 1992), *Hevea brasiliensis* (Gunasekaran and Krishnasamy, 1998); Jamun (Srimathi *et al.*, 2001) and arecanut (Raja, 2001).

6.3.2. Storage at sub imbibed condition

The principle of moist imbibed seed storage is based on the assumption that recalcitrant seeds are best maintained at moisture contents close that of their shedding moisture (Hong and Ellis, 1996). Johnston and Frolich (1957) reported successful storage of Avocado (*Persea americana* Mill.) seeds for several months by sand or sawdust preservation at 4.4 to 7.2°C and for eight months in dry peat at 5.5°C (Halma and Frolich, 1949). Seeds of *Trichilia dregeana,* T. *emetica, Podocarpus henkelii* and *Syzygium cuminii* were assessed for storability under sub-imbibed condition i.e. storage at relatively high water content but below the fully hydrated level, might prevent germination but would not be sufficient to be injurious to the seed. Eggers (2010) dried seeds of T. *dregeana,* T. *emetica,* P. *henkelii* and S. *cuminii* to various target moisture contents (which were determined for each species in the initial drying experiment) and then subjected to storage for 3-22 weeks at 6, 16 and 25°C (in sealed containers). Seeds of each species were also stored at the shedding water content to serve as control. The seeds were periodically removed for sampling and assessed for water content, germination, respiration, electrolyte leakage and microscopical features. Storage temperature appeared to affect viability of T. *emetica* and T. *dregeana* which displayed characteristics of chilling sensitivity. Storage at 6°C was detrimental (when compared with seeds stored under the same conditions at 16 and 25°C) regardless of whether the seeds were undried or partially dried prior to storage. The seeds of P. *henkelii* did not demonstrate chilling sensitivity, the viability not being compromised at 6°C compared with those seeds stored at 16 and 25°C. Jamun (*Syzygium cuminii)* seeds were not subject to storage at 6°C because previous work indicated that they would be chilling-sensitive (Srimathi *et al.,* 2001).

6.3.3. Partial dehydration

In partial dehydration, the recalcitrant seeds are dried to certain critical moisture content by air at a temperature of 20°C and then stored. Rubber seeds can be stored for one year with 50% germination if they are cleaned, soaked in 0.3% benlate, surface air dried and stored in perforated polythene begs at an ambient temperature of 25°C (Chin, 1989). Similarly, the cocoa seeds can be stored up to 24 weeks with more than 50% viability (Hor, 1984).

6.3.4. Effects of drying

Differential scanning colorimetry was used to study the relationship between drying rate, desiccation sensitivity and properties of water in homeohydrous (recalcitrant seeds) of *Landolphia kirkii*. Slow drying of intact seeds to axis moisture contents of approximately 0.9 - 0.7 g/g caused lethal damage while, flash drying of excised embryonic axes, permitted removal of water to approximately 0.3 g/g. The amount of non freezable water in embryonic axes (0.28 g water / g dry mass) did not change with drying rate and was similar to desiccation tolerant seeds. The results proved that the amount of non freezable water *per se* is not an important factor in desiccation sensitivity. However, flash drying that removed all freezable water damaged embryonic axes. Differences between desiccation sensitive and tolerant seeds occur at two levels **a.** tolerant seeds naturally lose freezable water, and sensitive seeds can lose this water without obvious damage only if it is removed very rapidly **b.** tolerant seeds can withstand the loss of a substantial proportion of non freezable water, whereas sensitive seeds are damaged if non-freezable water is removed (Pammenter *et al.,* 1991).

Leprince *et al.* (1999) studied the effect of desiccation in cotyledons and embryonic axes of *Castanea sativa* (Mill.) using Electron Spin Resonance spectroscopy technique having two nitrous oxide spin probes introduced into seed tissues to judge the cytoplasmic viscosity and membrane permeability. Whatever the drying rate involved viz., fast or slow drying, it had no effect on the rise in viscosity of axes or cotyledons. Cytoplasmic viscosity was same in both embryonic axis and cotyledon at 0.2 Poise until 1.6g water/g seed dry weight but increased exponentially thereafter. The embryonic axes were found to be more tolerant to drying than cotyledons. The cell membranes showed no damage and 50% viability was retained even when moisture content was reduced to 0.12 g/g. While, in cotyledons, the plasma membranes lost their integrity even at 0.6 g/g, thus embryonic axes and cotyledons had different response to drying rates. The metabolic rate also differed as was substantiated by reduced oxygen uptake in embryonic axes while it increased in cotyledons by 1 to 4 times even when reduced to 1.2 and gradually declined at 0.9 g/g with loss of membrane permeability and then rose to 2.8. This was again proved by reduction in cytochromes c and aa as measured using light spectroscopy which indicates irreparable damage to mitochondria and disruption of electron transport mechanism caused by desiccation. They concluded that the desiccation sensitivity in recalcitrant seed is due to the inability of the seeds to tone down their metabolism during desiccation and thus increasing the irreversible peroxidiative damage.

6.3.5. Low temperature storage

Seeds of *Calophyllum inophyllum* could be safely stored for one month in sealed polybags at temperature of 10-20°C. While in case of *Decalepis hamiltonii,* seeds collected during March-April were found to be mature with 83% germination which could be improved by soaking in hot water 60°C for 24 hrs. Low temperature storage was not suitable but ambient temperature storage (32±1°C) was successful. However, they were classified "intermediate" in storage behaviour, as storage in ambient conditions in sealed plastic pouches was not more than two months. In *Garcinia gumminguta,* shade drying was found suitable than sundrying. Seeds could be safely stored for one year at 20°C but did not show good response to pre storage treatments like application of protectants and moist sand storage and were classified as intermediate storage behaviour. Hence, storing *Garcinia* seeds in air tight pet bottles without seed treatment is recommended. While, Tang and Fu (1993) reported high seed viability when jackfruit seeds were stored for two years after washing and packing in polybags and stored at 15°C, Shylla Merlin and Palanisamy (2000) successfully stored the jack seeds at 10°C for 56 days. Ezumah (1986) on the other hand reported that cold storage (6-7°C) of neem (*Azadirachta indica*) gave poor results than at ambient temperatures of 26-28°C in contrast to Dhruv Kumar Misra *et al.* (1998) who suggested storing neem seeds at 4°C effective for 3 months. Seeds of jamun withstood storability at 10°C when preserved in 2% moist sand for 3 months (Srimathi *et al.,* 2001). Ban oak (*Quercus leucotrichophora*) seeds were best stored at 5°C in poly bags for 9 months (Bhardwaj *et al.,* 2001a), neem for 4 months (Selvakumar, 1999).

6.3.6. Use of abscisic acid in seed storage

Due to desiccation sensitivity of recalcitrant seeds, it is not possible to store them at ambient conditions as their metabolic activity is still high leading to loss of viability. Further, there is also the possibility of fungal invasion. Hence a metabolic inhibitor can help in arresting the metabolic activity.

Several references have proved that ABA can induce desiccation tolerance. Carrot somatic embryos treated with ABA can withstand desiccation tolerance and germinate later, which was not the case for untreated somatic embryos (Iida *et al.*, 1992). *Avicennia* though classified as recalcitrant has very low ABA during maturity along with other hormones (Farrant *et al.*, 1993). Since ABA can indirectly induce desiccation tolerance through biochemical mediation and the production of mRNAs and proteins (Bartels *et al.*, 1988; Mundy and Chua, 1988; Bruggink and Toorn, 1995) along with soluble carbohydrates content leading to desiccation tolerance (Tetteroo *et al.*, 1994).

Application of abscisic acid at 10^{-4} M to vermiculite resulted in moderate inhibition of germination and effectively improved the conversation in *Inga uruguensis* seeds (Barbedo and Cicero, 2000). ABA thus, could be effectively used in this regard as it is naturally present in seeds. However in viviparous seeds only 25 -50% of the normal concentration of ABA was found or the embryos were less sensitive to ABA and only high concentrations of ABA was found to inhibit germination (Bewley and Black,1985).

6.3.7. Hydropriming and storage

Hydropriming is a novel technique for imparting vigor to low vigor seeds. Various physiological enhancements like quenching of free radicals, activation of mRNA, invigoration of mitochondria have been attributed to priming. Vanangamudi *et al.* (2002) attempted this technique for treating three week old high and medium vigor neem drupes by drying them to 10% seed moisture and then hydration for 24 hrs followed by air drying. Germination per cent increased by 21% while seedling length, dry matter production of seedling significantly higher than control. While hydropriming followed by dry dressing with fungicide bavistin was detrimental. Hydropriming offers immense potential in alleviating the problems associated with high metabolic activity seen in recalcitrant seeds. More research needs to be done.

6.3.8. Controlled atmospheric storage

Due to the presence of high moisture and consequent high metabolic rate, emanation of toxic gases is imminent. These are to be removed or replaced. Providing more head space over stored recalcitrant seed is the first step in this direction. The recalcitrant seeds can be stored in controlled atmosphere of carbon dioxide or sealed containers. Cocoa seeds can be stored up to 45 days in carbon dioxide atmosphere (Villa, 1962) and durian seeds in sealed containers for about 32 days (Soepadmo and Eow, 1976).

Wills *et al.* (1981) stated that the wax formulations applied on fruits generally inhibit senescence. He suggested that the most successful storage method for cocoa could be storing the seeds within pods coated with paraffin wax. Similarly, Friend (1964) stored cocoa seeds within pods coated with paraffin wax. Chin (1989) stated that this type of controlled atmosphere storage would have much practical application in the storage of recalcitrant seeds. Raja *et al.* (2002 a) reported that the wax-coated arecanut seeds stored in gunny bag under ambient condition were successfully extended the storage life up to 50 days with 60% viability. While, uncoated seeds stored in ambient condition loses its complete viability within 24 days. The litchi seed retained with fruits treated with benomyl (0.05%) and wax emulsion (6%) and sealed in polythene bags maintained 42% viability for 24 days (Ray and Sharma, 1987). Shivarama Reddy (1987) found similar results in coffee seeds stored up to 300 days with 49% viability. The viability retained due to wax coating might be due to the reduced rate of respiration and moisture loss.

Coating had little effect on internal CO_2, O_2 and C_2H_4 levels but had significant effect in reducing water loss (Durand *et al.*, 1984; Hagenmaier and Baker, 1993; Banks *et al.*, 1997; Amarante, 1998). Controversially, coating of seeds with substances such as paraffin wax might be expected to restrict oxygen access to the seed and reduce storage life as evidenced in *Shorea javanica* (Umboh, 1987). Therefore ventilation of recalcitrant seed is essential to remove the excess toxic gases and to prevent anoxia (Tompsett, 1992). The acorns of *Quercus alba* can be stored more successfully if plastic bags with a wall thickness of only 0.04 mm are used to provide more aeration (Rink and Williams, 1984).

6.3.9. Cryopreservation and *in vitro* conservation

Reports over the past fifty years have shown that recalcitrant seeds can only be stored for short periods of few days or months to a year. Hence, long-term storage of recalcitrant seed is rapidly gaining momentum simultaneously with the development of *in vitro* technique (Roberts *et al.*, 1984).

Desiccation tolerance in recalcitrant seed increases during seed development on mother plant, but unlike in orthodox seeds, maturation drying to low moisture contents does not occur and fresh recalcitrant seeds have high levels of moisture content while shedding. Differences among various parts for seed moisture were also noticed and generally, storage tissues of recalcitrant seeds possessed lower moisture content than embryonic axis. Desiccation of excised embryos or embryonic axes paves way for conservation of recalcitrant embryos as it is possible to dry them to low moisture contents.

The recalcitrant embryos are tolerant to desiccation and low temperatures than whole seeds and together with their smaller size are amenable and practical for conservation (Chin, 1994). The strategy adopted for long-term genetic conservation of these recalcitrant seeds is to cryopreserve the embryos, which are more resistant to adverse conditions. Bajaj (1985) suggested that the germplasm of recalcitrant seeds could possibly be conserved through cryopreservation of their excised embryos. Cryopreservation of zygotic embryos and embryonic axes were reported by Normah *et al.* (1986); Pritchard and Prendergast (1986); Pence (1990, 1991); Chaudhury *et al.* (1991); Vertucci *et al.* (1991); Radhamani and Chandel (1992) and Chaudhury and Chandel, (1995).

Cryopreservation of embryonic axes is a reliable method for preservation for several mega seed crops (Boucaud *et al.* (1991); Pritchard and Prendergast (1986) and Pence (1990). Several cryopreservation protocols have been developed for a large number of species. Among the methods prescribed, pre-growth with cryoprotectants and slow freezing followed by encapsulation or dehydration were suggested for somatic embryos (Dereuddre *et al.*, 1991).

However, other methods like rapid freezing by immersion in LN or flash drying has also been developed for *Hevea brasiliensis* (Normah *et al.*, 1986); *Landolphia kirkii* (Berjak *et al.*, 1990) and *Nothapodytes nimmoniana* (Radha *et al.*, 2010) or flash drying followed by freezing at an intermediate temperatue (Vertucci *et al.*, 1991) or ultra rapid rate freezing (Wesley Smith *et al.*, 1992) have also been proposed.

However, embryonic axes was the preferred storage material as it was well differentiated and easily manageable and regeneration after LN preservation depends on protocol followed such as duration of desiccation, rate of drying, pretreatment with chemical protectants, pre freezing temperatures and cooling rates (Benson, 1994).

A brief tableau (Table 6.1.) of research in cryopreservation as documented by Engelman *et al.*, (1995) is provided below :

Table 6.1 : Attempts in cryopreservation of somatic and zygotic embryos (Engelman *et al.*,1995)

Species	Reference
Somatic embryos	
Citrus sinensis	Marin and Duran Vila,1988, Mann *et al.*,1993
Citrus arabica	Bertrand-Desbrunais *et al.*, 1988
Coffea canepora	Bertrand-Desbrunais *et al.*,1991; Hatanaka *et al.*,1994, Tessereau *et al.*,1994
Cucumis melo	Shimonishi *et al.*,1991
Daucus carota	Withers 1979; Tesssereau *et al.*,1994
Elaeis guineensis	Engelmann *et al.*,1995; Durnet *et al.*,1933
Juglans sp.	De Boucaud *et al.*,1994
Manihot esculenta	Sudarmonowati and Henshaw,1990
Xanthosoma sp.	Zandvoort,1987
Zygotic embryos	
Aesculus hypocastanea	Pence,1990
Arachis hypogaea	Runthala *et al.*,1993
Araucaria exelsa	Pritchard and Prendergast,1986
Artocarpus heterophyllus	Krishnapillay,1989
Brassica napus	Withers, 1982
Carva sp.	Pence,1990
Camellia sinensis	Chaudhury *et al.*,1991
Castanea sp.	Pence,1990
Citrus sinensis	Radhamani and Chandel,1992
Cocos nucifera	Chin *et al.*,1989; Assy-Bah and Engelmann, 1992
Coffea sp.	Normah and Vengadasalam,1992, Abdelnour *et al.*,1992
Coryllus avellana	Gonzales-Benito and Perez,1994; Normah *et al.*,1986, Reed *et al.*,1994
Eleaeis guineensis	Grout *et al.*,1983
Hevea brasiliensis	Norma *et al.*,1986
Jugalans sp.	Pence,1990
Landolpha kirkii	Vertucci *et al.*,1991
Musa sp.	Abelnour *et al.*,1992
Theobroma cacao	Pence 1991

Successful cryopreservation is possible when excised embryonic axes survive desiccation below the threshold freezable moisture content of about 22-27% for *Landolphia kirkii*, 30-33% for *Acer saccharinum*,

Nephelium lappaceum, Durio zibehtinus and *Artocarpus heterophyllus* (Ellis and Hong, 1996), and 19.6% for *Nothapodytes nimmoniana* (Radha *et al.,* 2010) below which value there is no freezable water for ice formation. Embryonic axes of *Aesculus hippocastanum* can be desiccated upto 12% moisture and subsequently cryo preserved. *Dimocarpus longan* embryos could be successfully dried to 18% and stored for 24 hr in LN (Ellis and Hong, 2006).

Method and drying rate are the most important factors that determine the storage longevity of cryopreserved embryos. The process involved excision, cleaning and sterilizing the seed using 0.01% mercuric chloride (w/v) for 5-10 min followed by 5-6 rinses in sterile distilled water and then decoating the seeds for dissecting the embryos and drying them in laminar flow chamber for various durations. Almond (*Prunus amygdalus*) embryonic axis when desiccated between 10-4% moisture could be preserved in liquid nitrogen by immersing embryonic axes placed in cryotubes for 24 hr at -196°C and thawed rapidly by plunging cryotubes in a water bath maintained at 38°C for 1-2 minutes and the axes were immediately transferred to culture medium containing Murashige and Skoogs nutrient content and maintained under light for 12 hr (25-50$\mu Em^{-2} s^{-1}$). Survival was recorded for few weeks to months depending on the type of species studied and assessed as the embryonic axes turning green or exhibited growth as envisioned by seedling development, shoot growth and root growth. The highest survival was at 7% moisture when normal healthy shoots and roots emerged from 66% of surviving axes. Desiccation below 7% moisture caused a decline in viability and regrowth of embryo (Table 6.2.).

Table 6.2 : Survival of almond embryonic axes estimated prior to and after cryopreservation

Hours of desiccation	Moisture content (% FWB)	Before cryopreservation Viability %	Rate of regrowth $	Rate of regrowth #	After cryopreservation Viability %	Rate of regrowth $	Rate of regrowth #
0	21.9	100	100	80	0	0	0
1	10.8	100	100	85	80	60.3	34.2
2	8.7	100	100	80	75	66.6	33.3
3	7.2	100	100	80	100	100	66.6
4	5.2	100	83	60	50	16.6	16.6
5	4.0	62	62	50	40	40	0

$ Shoot emergence per cent of surviving axes with normal shoots 3 months after culture
# Root emergence per cent of surviving axes with normal roots 3 months after culture

Successful cryopreservation, at very low temperature demands that excised embryos must be dried to suitably low moisture content to avoid ice formation by ultra low temperature (Hor *et al.*, 1990). A number of species of recalcitrant type, both temperate and tropical clones have been known to survive after cryopreservation (Poulsen, 1992), which includes *Juglans, Carya, Fagus, Corylus, Castanea* (Pence, 1990), *Landolphia kirkii* (Pammenter *et al.*, 1991), litchi (Fu *et al.*, 1994), coffee (Engelmann *et al.*, 1995), *Aesculus* spp. (Pence, 1990), *Araucaria hunsteinii* (Pritchard and Prendergast, 1986), *Artocarpus heterophyllus* (Krishnapillay,1989), *Castanea sativa* (Pence,1990, Palanisamy and Pritchard, 1999), *Coffea liberica* (Normah and Vengadasalam, 1992), *Cocos nucifera* (Chin *et al.*, 1989), *Dimocarpus longan* (Fu *et al.*, 1990), *Havea brasilensis* (Normah *et al.*, 1986), *Quercus* spp. (Pence, 1990), *Theobroma cacao* (Pence,1991), *Areca* (Raja *et al.*, 2004) and *Nothapodytes nimmoniana* (Radha *et al.*, 2010).

Pammenter *et al.* (1991) working with *Landolphia kirkii,* suggested flash freezing of seeds to 0.7-0.9 g/g moisture content compared to slow drying. Flash drying did not disturb non-freezable water in the embryonic axes but removed all freezable water that damaged embryonic axes. The cryopreservation techniques involved had their drawbacks. Overall membrane damage with high degree of vacuolation, withdrawal of cytoplasm, disruption of nucleus, mitochondria, endoplasmic reticulum and golgi bodies was noticed in embryonic axes of cocoa, tea and jackfruit. Withdrawal or disruption of plasma lemma was noticed in cocoa.

Another feature was the disruption of storage organelles like starch granules. This indicated even if they survived the freeze treatment, their survival was limited (Chandel *et al.*, 1995). The damage to viability of axes exposed to liquid nitrogen at high moisture content results from the presence of freezable water in tissues, which contributed to the formation of ice crystals in the intercellular space within the axes (Zewdie and Ellis, 1991; Assy–Bah and Engelmann, 1992; Fu *et al.*, 1993). Similarly desiccation of axes if progressed beyond certain moisture content, leads to reduction in survival. It is probably due to severe dehydration injury (Assy-Bah and Engelmann, 1992; Engelmann *et al.*, 1995; Berjak and Dumet, 1996). Radha *et al.*, (2010) working with *Nothapodytes nimmoinana* embryos recorded damage to plumule caused by dehydration rather than radicle as the cause for death in cryo-freezed embryos.

A method to over come freezing damage was proposed by Koster *et al.* (2003) that freezing pea protoplasts led to decreasing in viability as desiccation damage increased. The protoplast isolated from radicles, 12-18 upon embryo desiccation were able to survive than when extracted after 24-26 hrs, when the embryos became desiccation sensitive. Although intact epicotyls are generally more desiccation tolerant than radicles, protoplasts isolated separately from epicotyls and radicles did not differ in their tolerance to desiccation. Protoplast survival extracted from embryos dehydrated by immersing in 500 mM sucrose/raffinose or 700 mM sucrose behaved similarly. Higher concentration of sugar solution made the protoplast tolerant to desiccation. The experiment proved that desiccation induced by strong sugar solution improved desiccation tolerance.

Another indicator of successful cryopreservation is selection of embryos that were mature, because susceptibility to freezing depended on physiological maturity of seed and the extent of fruit maturity. Research carried out using coffee seeds (C. *arabica*) extracted from two different fruit colour stages viz., green and red revealed that even though the desiccation period ensuring highest survival period was similar for both embryos, higher percentage (96%) of mature embryos withstood the rigors of cryopreservation while only 50% of immature were capable of doing so (Engelmann *et al.*,1995) (Table 6.3.).

Table 6.3 : Cryopreservation of embryonic axes extracted from various maturity stages of coffee

Desiccation (hrs)	Survival (%)			
	Immature embryos		Mature embryos	
	-LN	+LN	-LN	+LN
0.0	100	0	100	0
0.5	80	50	100	96
1.0	53	34	46	42
1.5	25	14	28	19
2.0	6	0	8	0

These embryos could be revived by placing them on a recovery medium containing GA (100 mg/L) resulting in 83% survival. Studies by Hong and Ellis (1995) also revealed difference among species for desiccation sensitivity. Similar observations were noted in coconut also where embryos extracted from 7-8 month old nuts had limited growth compared to 11-12 months old. The cause was ascertained as the inadequate recovery conditions for immature embryos (Assy-Bah, 1992).

Partial desiccation of embryos prior to cryopreservation is essential for survival after cryopreservation. Oil palm embryos desiccated to 0.3g water/g dry weight and frozen in LN showed high survival rates, but caused irreversible damage to 30% per cent of them. This could be due to the modifications induced such as non-development of haustoria as noticed in *Veitchia, Howea* and coconut embryos. Working with embryos of *Inga* spp., Pritchard *et al.* (1995) found that level of sensitivity was independent of species, seed lot, desiccation rate and was influenced by drying environment and physical characteristics of embryos. This behaviour was due to the dual characteristics of desiccation intolerance and lack of protective mechanism provided by presence of soluble sugars.

6.4. Preservation as Synthetic Seeds

Synthetic seeds (Redenbaugh *et al.*, 1986) has a huge potential as it will be most suitable technique to envelope a recalcitrant embryo. Since synthetic seed comprises of an embryo separated from its accessory structures and encapsulated in a hydrated gel capsule that will permit natural development of the embryo. Studies have shown that excised encapsulated embryos are able to withstand lower temperature and store longer than intact seeds. Sudhakara *et al.* (2000) tried a novel experiment of placing excised cocoa embryos in 4% sodium alginate and dropping them into 75 mM Calcium chloride solution for 30 min to form a gelatinized bead. These beads known as synthetic seeds were stored at 10°C for 25 days in wet cotton and estimated their viability potential. The conversion rate declined from initial level of 97.3% to 71% after 25 days. But it took 45 days to complete the germination when it was just 24hrs for just encapsulated embryos. Further the emergence of radicle and completion of germination was faster in synthetic seeds than in fresh seed intact embryos highlighting the role of encapsulated excised embryos. Encapsulated embryos of *Hopea parviflora* also showed better storage potential at 10°C for 4 weeks compared to intact seeds.

Abscisic acid, a potential germination inhibitor can be used induce dormancy in synthetic seeds. It has been extensively used to harden embryos of carrot (Kitto and Janick, 1985) and celery (Kim and Janick, 1989), thus surviving desiccation. This theory was put into practice by Sunilkumar *et al.* (2000) by creating synthetic seeds encapsulated using *Hopea parviflora* (a recalcitrant *Dipterocarpaceae* timber species) embryos extracted from seeds soaked overnight in 1-3 mg/L ABA showed that

2-3 mg/L ABA treated embryos had higher ability to withstand low temperature storage and also germinate faster (Table 6.4.). Lower concentration did not prolong the viability while higher concentration was inhibitory as envisaged by delayed germination. This can be nullified by addition of cytokinin (Sunilkumar *et al.*, 2000).

Table 6.4 : Effect of ABA treated and encapsulated embryos of *Hopea parviflora*. (values in parenthesis indicate days taken to initiate germination).

Conc. of ABA mg/L	Storage of synthetic seeds (wks)				
	0	1	2	3	4
0	85 (6)	85 (3)	90 (5)	75 (6)	75 (5)
1	100 (2)	85 (3)	65 (4)	0	0
2	100 (2)	85 (3)	84 (4)	50 (6)	85 (3)
3	60 (3)	50 (4)	35 (5)	35 (7)	85 (5)

An attempt by Pammenter *et al.* (2002) to encapsulate *Avecennia marina* seeds in sodium alginate and polymerized in calcium chloride solution akin to synthetic seeds. Encapsulated seeds had higher degree of storability but time for first germination and rate of protein synthesis did not differ between naked and encapsulated seeds. Fungal contamination posed a major problem for hydrated storage of *Avecennia* seeds, though fungal incidence was low in encapsulated seeds.

□□□

Chapter 7

Role of Vigor Tests

Germination rate may provide the first indication of desiccation stress in recalcitrant seeds; thus mango exhibits a reduced vigour index before there is any noticeable fall in germination *per se* (Fu *et al.*, 1990). Dhruv Kumar Misra *et al.* (1998) suggested top of paper as the best method for assessing germination for neem followed by vermiculite and sand. Removal of endocarp facilitated early emergence in neem though no change in germination was recorded. The freshly harvested clove seeds are the best suited for sowing in order to get maximum germination and vigorous seedlings (Sabale *et al.*, 1992). Horizontal placement of seeds with micropylar end facing sideways was the best for getting optimum and vigorous seedlings in nutmeg, clove and rubber (Gunasekaran, 1997). Normah and Chin (1991) observed decreased germination, seedling height, seedling dry weight and respiration rate with increased storage of rubber seeds. Seedling length proved to be a good index of seed vigor in jamun (*Eugenia jambolana*) (Srimathi *et al.*, 2001). Raja *et al.* (2001a) observed decreased viability and seedling vigour during the desiccation of recalcitrant avocado seeds. In another study Anderson (1970) stated that reduction in seedling growth precede or accompany loss of germinability and need not occur in every case of seed deterioration.

Vigor indices obtained using seed germination values viz., germination capacity, germination energy and germination value were found to be better correlated with germination per cent for Ban oak (Bhardwaj *et al.*, 2001a). Similar tests were also used to indicate the poor storability of *Ulmus laeigata* Royle seeds (Bhardwaj *et al.*, 2001b) and neem (Dhruv Kumar Misra *et al.*, 1998). Gunasekaran (1997) reported increased values for electrical conductivity (EC), pH, free amino acid and reducing sugars for stored seeds of nutmeg, clove and rubber. While Selvakumar (1999) reported reduced vigor index, protein and oil content of stored neem drupes while EC values increased.

Beewar *et al.* (1982) found that as embryos of silver maple were desiccated from 45 to 35%, the leakage from them increased while germination decreased from 97- 5%. In case of areca palm (*Chrysalidocarpus lutenscens*) embryos showed increase in leakage and steady state leakage when dehydrated from 84 to 53%.

Bharathi (1999) suggested removal of neem endocarp followed by preconditioning in cold water for 24 hrs for removing seed coat and then soaked in 0.2% tetrazolium solution for assessing the viability using Tetrazolium. The seeds have to be kept in dark for 1 hr at 40°C for visualizing the Tz pattern. In case of coffee, experiments with cv. Tafaricala endorsed the removal of parchment membrane by preconditioning in water for 20 hrs, slicing the endosperm and removing the embryo for soaking in 0.5% Tz solution and placing in dark for 6 hr at 30°C for uniform staining. The TTZ absorbance was also significantly reduced. They suggested that, membranes of desiccation sensitive seeds were damaged by dehydration below critical moisture content which was 40% for silver maple and 55% for areca palm embryos and that membrane damage is the cause for viability loss.

Other tests like paper piercing test, standard germination test, 100 seed weight, dehydrogenase activity, peroxidase test and EC were suited to assess the vigor for neem (Bharathi, 1999).

A special test known as sprout test was proposed by Ponnusamy *et al.* (2007) where coffee seeds soaked in water (100 seeds in 100 ml) and germinated. Decoated seeds sprouted in 24 hrs while seeds with parchment membrane intact took 96 hrs, indicating the inhibitory role of the membrane.

It has been observed in several cases that recalcitrant seeds take a long time to germinate, thus confounding the researcher about its viability. Jamun seeds require 60 days to germinate. In such cases, vigor tests can be used as these are based on physical, physiological and biochemical characteristics of the seeds and are inexpensive, uncomplicated, rapid, objective, reproducible and highly correlated with nursery or field emergence. Srimathi *et al.* (2001) reported higher degree of correlation with EC values of stored jamun seeds with germination three months into storage. Further, germination tests do not reflect the progress of deterioration. Cibele *et al.* (2003) while working with King Palm seeds reported, that as desiccation progressed, the electrical conductivity values increased and reached maximum when no germination occurred. Similarly, the germination also reduced from initial value of 67.5% to 0.0% when dried for 156 hrs (Table 7.1.).

Table 7.1 : Vigor indices of King Palm seeds

Seed moisture (%)	First count (%)	Abnormal seedlings (%)	Dead seeds (%)	Total germination (%)	EC mS/cm/g
47.5	61.2	11.2	21.2	67.5	14.78
38.8	53.7	5.0	28.7	66.2	19.30
35.4	52.5	5.0	30.0	65.0	21.40
34.6	46.2	3.7	38.7	57.5	22.10
31.5	46.2	20.0	27.5	52.5	21.33
30.1	30.0	17.5	41.2	41.2	22.50
28.2	21.2	13.7	48.7	37.5	23.78
26.6	17.5	1.2	77.5	21.2	23.40
24.8	10.0	1.2	71.2	27.5	25.43
23.9	12.5	3.7	76.2	20.0	24.85
22.0	5.0	0.0	85.0	15.0	26.18
20.0	1.2	0.0	86.2	13.7	30.73
17.8	0.0	1.2	92.5	6.2	30.98
15.1	0.0	0.0	100.0	0.0	46.70

Cibele *et al.* (2003) also found that the EC values were negatively correlated with moisture content, indicating the essentiality of using EC test as a good indicator of seed viability (Table 7.2.).

Table 7.2 : Correlation coefficient for various vigor test also proved the ability of EC test to predict the fall in viability of King Palm

Traits	First count germination	Dead seed	Total germination	Abnormal seedling	Plumule length	Radicle length	EC
Seed moisture	0.81***	-0.86***	0.87***	0.42***	0.75***	0.83***	-0.78***
First count		-0.91***	0.90***	0.51***	0.65***	0.81***	-0.57***
Total germination			-0.97***	-0.63***	-0.67***	-0.78***	0.68***
Abnormal seedling				0.43***	0.71***	0.80***	-0.68***
Plumule length					0.23 NS	0.36**	-0.36***
Radicle length						0.86***	-0.76***
EC							-0.70***

In some cases, it may not be possible to measure the EC values of whole seeds especially for mango, coconut etc. Under such circumstances measurement of conductivity and lipid peroxidation values of fresh and desiccated embryonic axes could be attempted. Studies conducted to determine the EC values of embryonic axes of tea, cocoa, jackfruit & French bean revealed EC was the best indicator of viability loss, though lipid peroxidation was also increased upon desiccation. But as measurement of conductivity is more economical it can be suggested as the best (Chandel *et al.*, 1995) (Table 7.3.).

Table 7.3 : Conductivity and Lipid peroxidation values of embryonic axes subjected to desiccation and immersion into LN

Crop	Treatments	Moisture (%)	Conductivity $\mu Scm^{-1} gd.wt^{-1}$	Lipid peroxidation TBA rp g $d.wt^{-1}$
Tea	Fresh	56.1	988.6	0.079
	4 H Desiccation	13.0	1017.2	0.070
	4 H D + LN	13.0	1080.4	0.080
Cocoa	Fresh	56.0	572.0	0.202
	4 H Desiccation	47.0	1218.2	0.235
	4 H D + LN	47.0	1855.5	0.940
Jackfruit	Fresh	62.0	2360.0	0.031
	4 H Desiccation	14.2	2407.0	0.027
	4 H D + LN	14.2	2917.0	0.060
French bean	Fresh	11.0	61.6	0.010
	4 H Desiccation	10.0	66.0	0.011
	4 H D + LN	10.0	95.7	0.014

The seed vigour was more closely related to the integrity of the protein synthesizing system than to the protein content of the seed (Abdul-Baki, 1980). Szezotka (1975) reported that both decreased protein synthesis and oxygen uptake are parallel to the decline in germination of *Quercus robur* and *Q. borealis.* Also protein synthesis is essential for germination to be completed and for radicle emergence to occur (Bewley and Black, 1983). Nautiyal *et al.* (1985) noticed lower amount of soluble proteins in non-viable seeds of *Shorea robusta* than in viable seeds.

A higher value of sinkers (%) was noticed in jamun seeds that had high germination potential into three months of storage. Since, seeds that sink when placed in water would possess higher moisture, thus retain viability in case of recalcitrant seeds, Srimathi *et al.* (2001) used this as an estimate of seed vigor in jamun.

Bailey *et al.* (1962) observed the presence of isovaleraldehyde, isobutyraldehyde, propionaldehyde, methanol, acetaldehyde, methyl acetate and diactyl, all volatile aldehydes, in deteriorating cocoa seed. Wilson and Mc Donald (1986) suggested that these deleterious gases arise from deteriorating seeds which were products of lipid peroxidation. Woodstock and Taylorson (1981) found growth inhibitory substances being released from low vigor seeds. Later Basu *et al.* (1990) developed the bioassay test which has been propounded as the most sensitive test that reflects inhibitory effects of the minute quantities of volatile aldehydes emanating from a dying seed over a vigorous and quick germinating seeds of jute. Bhattacharyya and Basu (1992) observed higher levels of volatile aldehydes in deteriorating jackfruit seeds that were inhibitory to growth. Arecanut dehusked drupes in various stages of storage retrieved at four days interval were used for in bioassay test using high vigor pearmillet seeds. The pearlmillet seeds were substantially affected by stored areacanut seeds due to the deleterious gases emanation (Raja *et al.*, 2004) (Table 7.4). Bioassay test was also successfully demonstrated to show the presence of inhibitory substances in coffee seed parchment (Ponnusamy *et al.*, 2007).

Table 7.4 : Bioassay of Areca seeds retrieved from storage on pearlmillet seeds.

Storage of Areca seeds (days)	Seedling attributes of pearl millet used in bioassay test			
	Germination (%)	Shoot length (cm)	Root length (cm)	Seedling dry weight (mg/10 seedlings)
Blank	89	13.0	15.4	63
0	77	12.7	13.1	60
4	72	12.1	11.9	54
8	70	11.2	11.6	55
12	66	11.0	10.5	54
16	65	10.6	9.1	53
20	58	10.6	8.1	50
24	56	8.9	6.9	50

X-ray analysis is another potent technique that can be successfully used to analyze recalcitrant seeds. Kamra (1976) had conducted X-ray radiographic analysis of 165 species from 9 tropical countries and suggested that radiographic analysis can be used to detect shrunken, shriveled embryos in woody endocarp species along with insect and mechanical damage. It can also be used to count the number of empty or filled chambers in hard drupes and the extent of embryo growth in recalcitrant seeds. Seed maturity and morphological maturity of embryos can be seen using x rays without damaging or pricing open the seeds, thus it becomes an veritable tool in case to those seeds which are rare or in few numbers (Jones, 1976). Contrast X-ray that uses seeds imbibed with contrast chemicals viz., barium chloride, ronographin, visotrast were also suggested as the best method to visualize damaged embryos (Simak, 1957). Using 16 hr preimbibed seeds for soaking in 20% $BaCl_2$ solution for 1 hr was the best for contrast X-ray study of Tea and Coffee, while it took 2hr soaking in $BaCl_2$ for rubber seeds (Banerjee *et al.*, 1973). The suggested procedures to be followed for mango were Voltage : 15kV; Current 2mA; FFD 25cm; Duration of exposure 3 sec. (Banerjee *et al.*, 1973). Xeroradiography was developed to produce a radiographic image in positive or negative mode for better visualization. A positively charged plate was used instead of X ray film. Such images are known as xerograms and were utilized by Vozzo (1988) to study the delicate features of black walnut (*Juglans nigra*). X-ray analysis is thus a non destructive tool. X-ray machines all already been utilized in motor vehicle industries to study the stress facture in metal parts. Such machines can be used to evaluate recalcitrant seeds prior to storage or evaluation, thus saving the cost of storing unviable, damaged and insect infected seed.

□□□

Chapter 8

Conclusion

Recalicitrant seeds are not an aberrant of nature, but an essential component of natural ecosystem. It is the intervention of humans that has brought their behavior to light thus posing challenges to collection, classification and storage. Research in recalcitrant seed behavior has been going on for several decades and will continue to go on for few more decades. Most of the research has been done in areas that are not familiar to recalcitrant seed, especially in hot humid tropics where several of the species, even cultivated crops (mango, durian, jamun, litchi, curry leaf, avocado, coconut, jackfruit) are raised commercially. Further intra and interseasonal variability across species as well as within individual species confounds the problem further with location playing a significant role in dermination of recalcitrant behaviour in individual species (Berjak and Pammenter, 2008). Taxonomic classification hence does not provide a basis for recalcitrant, rather it is a derived trait among species and apprears more in species of mesic tropics, especially among rainforest species and in cooler, drier and seasonal habitats.

Drying of recalcitrant seed still pose a major problem and is to be cataloged and prescribed. Unless protocols for drying are provided, world may continue to loose most of the collected recalcitrant seeds of commercial value and germplam needs. Storage of recalcitrant seed is another grey area, since most of the technology available sofar gurantees storage for a few days / weeks. Storage of embryonic axes may solve the problem to an extent, but the interaction of embryo and cotyledon has not been researched extensively. Till then, ways and means have to be developed to store the seeds.

The greatest impediment to storage of recalcitrant seed is that they are continuously metabolically active. This generates lot of free radicals and also promotes cellular activity. Method to slow the metabolism through hormonal-chemical cornucopia has to be researched.

Identification of candidate plus trees has been well documented for timber species of temperate regions and to some extent for commercially important crops. But such criteria are yet to be provided for several others. Until, evaluation of candidate plus tree is provided, storage of poor and vulnerable types will be done, which will entail loss economically and genetically. Promotion of seed processing labs or mobile assessment and processing labs close to areas of collection will improve collection and preparation of seed for transport. Fungal load is a major problem that domintes recalcitrant seed storage. Methods to overcome fungal infection or even prevent it have to be explored.

Many aspects of seed sampling, testing which is the norm for several agricultural and horticultural crops are not available or difficult to apply for recalcitrant seeds. Though several seed testing procedures have been available for timber and forest trees of commercial importance, no such attempt has been made for recalcitrant species.

Development of vigor test is another area of seed technological importance. Research has already pointedout the conformatory attributes of electrical conductivity test for assessing seed leacheate. X ray radiography can improve the collection and promotion of insect free seeds. Tetrazolium staining can be used to select good quality embryonic axes for cryopreservation. These and several other vigor tests can be successfully used to promote and store vigorous seeds.

As seeds range in their desiccation ability from desiccation tolerance - desiccation tolerance but chilling sensitive to desiccation and chilling sensitive thus a continuum, research in recalcitrant seed is also a continuum.

Last but not the least, research in any field is a continuing process and what is known is very little. As ancient Tamil poetess Ov-vyyar quoted "what we know is just a handful, what we do not know is as vast as the ocean".

□□□

References

Abdul-Baki, A.A. 1980. Biochemical aspects of seed vigour. *Hort. Sci.*, 15: 765-771.

Amarante, C. V. T. 1998. Gas exchanges, ripening behaviour and post harvest quality of coated pears. Ph.D. Dissertation, Massey Univ., Palmerston North, New Zealand.

Anderson, J. D. 1970. Physiological and biochemical differences in deteriorating barley seed. *Crop Sci.*, 10: 36-39.

Ang, B. B. 1977. Problems of rubber seed storage. In: Seed Technology in the Tropics. (Eds. H.F. Chin, I.C. Enoch and R.M. Raja). Harun Univ. Pertanian Malaysia. pp. 117-112.

Assy-Bah, B. 1992. Utilization de la culture in vitro d'embryons zygotiques pour la collecte et al conservation des resources genetiques du cocotier (*Cocos nucifera* L).These d'Universite, Uriversite Paris 6:146p.

Assy-Bah, B. and F. Engelmann. 1992. Cryopreservation of mature embryos of coconut (*Cocos nucifera* L.) and subsequent regeneration of plantlets. *Cryo Letters*, 13: 117-126.

Bacchi, O. 1961. Estudos sobre a conservação de sementes: IX - Ingá. Bragantia, 20: p.805-814.

Bailey, S. D., D. G. Mitchell, M. L. Bazinet and C. J. Weurman. 1962. Studies on the volatile components of different varieties of cocoa beans. *J. Food Sci.*, 27: 165-170.

Bajaj, Y. P. S. 1985. Cryopreservation of embryos. In: Cryopreservation of Plant Cells and Organs. (Ed. K.K. Kartha). CRC Press, Florida. pp. 228-242.

Banerjee, S. K., Amir Singh, M. M. Verma and K. Verma.1973. X ray analysis of seeds. *ICAR Tech. Bull.* (Agric.) 47 : 33 p.

Banks, N. H., Q. Cheng, S. E. Nicholson, A. M. Kingsley and P. B. Jeffery. 1997. Variation with temperature in effects of surface coatings on gas exchange of apples. Invited Presentation to Internatl. Cong. for Plastics in Agric., 9-15 March, Tel Aviv, Israel.

Barbedo, C. J. and D. A. C. Bilia. 1998. Evolution of research on recalcitrant seeds. *Scientia Agricola* 55 (*Special Issue*).

Barbedo, C. J. and S. M. Cicero. 2000. Effects of initial quality, low temperature and ABA on the storage of seeds of *Inga uruguensis*, a tropical speices with recalcitrant seeds. *Seed Sci. & Technol.*, 28: 793-808.

Bartels, D. M. Singh and F. Salamini. 1988. Onset of desiccation tolerance during development of the barley embryo. *Planta*, 175 : 485-492.

Basu, R. N., P. Kuchlan and K. Sur .1990. Bioassay of seed vigour. *Ind. J Exptl. Biol.* 28 : 166 - 170.

Beewar, M. R., P. C. Stanwood and E. E. Roos. 1982. Dehydration effects on imbitional leakage from desiccation sensitive seeds. *Plant Physiol* 69:1132-1135.

Benson, E. E. 1994. Cryopreservation. In : R. A. Dixon and R. A. Gonzales. (Eds.). Plant Cell Culture : A practical approach. IRL Press, Oxford. Pp: 147-167.

Berjak, P, J. M. Farrant, D. J. Mycock and N. W. Pammeter. 1990. Recalcitrant seeds : the enigma of their desiccation sensitivity. *Seed Sci. & Technol.* 18: 297-310

Berjak, P. 1989. Storage behaviour of seeds of *Hevea brasiliensis*. *J. Natural Rubber Res.*, 4 (3): 195-203.

Berjak, P. 1995. The role of microorganisms in deterioration during storage of recalcitrant and intermediate seeds. In: Intermediate / Recalcitrant Tropical Forest Tree seeds. (Eds. A.S. Ouedraogo, K. Poulsen and F. Stubsgaard). Proc. Workshop on Improved Methods for Handling and Storage of Intermediate/Recalcitrant Tropical Forest Tree Seeds, 8-10 June, Humlebaek, Denmark.: pp. 121-126.

Berjak, P. and D. Dumet. 1996. Cryopreservation of seeds and isolated embryonic axes of neem (*Azadirachta indica*). *Cryo Letters*, 17: 99-104.

Berjak, P. and N. W. Pammenter. 1994. Recalcitrant is not an all-or-nothing situation. *Seed Sci. Res.*, 4 : 263-264.

Berjak, P. and N. W. Pammenter. 1996. Recalcitrant (Desiccation-sensitive) seeds. In. Innovations in tropical tree seed technology. Proc. Of the IUFRO symposium of the Project Group P.2.04.00 'seed problems' (Ed. Kirsten Olesen), Danida, Denmark.: Pp14-29.

Berjak, P. and N. W. Pammenter. 2000. Orthodox and Recalcitrant Seeds. In. Tropical Tree Seed Manual (Ed.). Vozzo, J.A., USDA Forest Service. Pp:137-147.

Berjak, P. and N. W. Pammenter.2008. From *Avicennia* to *Zizania* : Seed recalcitrance in prespective – a review. *Ann. of Bot.* 101 : 213-228.

Berjak, P., K. J. Bradford, D. A. Kovach and N. W. Pammenter. 1994. Differential effects of temperature on ultrastructural responses to dehydration in seeds of *Zizania palustris*. *Seed Sci. Res.*, 4: 111-121.

Berjak, P., M. Dini and N. W. Pammenter. 1984. Possible mechanisms underlying the differing dehydration responses in recalcitrant and orthodox seeds. *Seed Sci. & Technol.*, 12 : 365-384.

Bernard, R. L. 1950. Seed storage trials in sawdust. Malaysian Forester, : 163-165.

Bewley, J. D. and M. Black. 1983. Physiology and Biochemistry of Seeds in Relation to Germination, Berlin Springer-Verlag. : Pp. 306.

Bewley, J. D. and M. Black. 1985. Seeds. Physiology of development and germination. New York : Plenum Press : 45-52.

Bharathi, A. 1999. Studies on handling, management and storage of need seed (Azadirachta indica A. Juss). Ph.D. Thesis. Dept. of Seed Science and Technology, TNAU, Coimbatore.

Bhardwaj, S. D., S. Sharma, S. K. Dutta and R. Banyal. 2001. Storage studies on Acorns of Ban Oak (*Quercus leutrichophora*). *Seed Res.* 29(1):28-33.

Bhardwaj, S. D., Pankat Panwar and B. S. Kanwar. 2001a. Effect of containers and temperature on longevity of *Ulmus laevigata* Royle seeds. *Seed Res.* 29(1):34-37.

Bhattacharyya, A. K. and R. N. Basu. 1992. Retention of vigour and viability of jack fruit (*Artocarpus heterophyllus* Lam.) seed. *Indian Agric.*, 36(2): 65-74.

Blackman, S. A., S. A. Obendrof and A. C. Leopold. 1992. Maturation proteins and sugars in desiccation tolerance of developing soybean seeds. *Plant Physiol.*, 100 : 225-230.

Blackman, S. A., S. H. Wettlaufer, R. L. Obendorf and A. C. Leopold. 1991. Maturation proteins associated with desiccation tolerance in soybean. *Plant Physiol.*, 96:868-874.

Bochicchio, A., C. Vazzana, A. Raschi, D. Bartels and F. Salamini. 1988. Effect of desiccation on isolated embryos of maize. Onset of desiccation tolerance during development. *Agronomie*, 8 :29-36.

Bonner, F.T. 1995. Commercial seed supply of recalcitrant and intermediate seed: Present solutions to the storage problem. In: Intermediate / Recalcitrant Tropical Forest Tree Seeds. (Eds. A.S. Ouedraogo, K. Poulsen and F. Stubsgaard). Proc. Workshop on Improved Methods for Handling and Storage of Intermediate / Recalcitrant Tropical Forest Tree Seeds, 8-10 June, Humlebaek, Denmark. pp. 27-33.

Bonner, F. T., 1996. Storage of temperate recalcitrant forest tree seeds. In. Innovations in tropical tree seed technology. Proc. Of the IUFRO symposium of the Project Group P.2.04.00 'seed problems' (Ed. Kirsten Olesen), Danida, Denmark.: Pp 30-36.

Boroughs, H. and J. R. Hunter. 1963. The effect of temperature on the germination of cocoa seeds. *Proc. Amer. Soc. Hort. Sci.*, 82: 222.

Boucaud, M. D., M. Brison, C. Ledoux, E. Germain and A. Lutz. 1991. *Cryoletters* 12 : 163-166.

Bradford, K. J. and P. M. Chandler. 1992. Expression of dehydrin-like proteins in embryos and seedlings of *Zizania palustris* and *Oryza sativa* during dehydration. *Plant Physiol.*, 99: 488-494.

Bruggink, T. and P. Toorn Van Der. 1995. Induction of desiccation tolerance in germinated seeds. *Seed Sci. Res.*, 5 : 1-4.

Brunori, A. 1967. A relationship between DNA synthesis and water content during ripening of *Vicia faba* seeds. *Caryologia* 20 : 333-338.

Castro, Y. G. P. and H. P. Krug. 1951. Experiências sobre germinação e conservação de sementes de *Inga edulis*, espécie usada em sombreamento em cafeeiros. *Ciência e Cultura*, 3 : 263-264.

Chacko, K. E. and R. N. Singh. 1971. Studies on the longevity of papaya, phalsa, guava and mango seeds. *Proc. Internatl. Seed Test. Assoc.*, 36(1): 147-156.

Chaitanya, K. S. K. and S. C. Naithani. 1994. Role of superoxide, lipid peroxidase and superoxide dismutase in membrane perturbation during loss of viability in seeds of *Shorea robusta* Gaertn. *F. New Phytol.*, 126: 623-627.

Chandel, P. S., J. Rekha Chaudhury, J. Radhamani and S. K. Malik. 1995. *Annals of Botany* (76) 5:443-450.

Chaudhury, R., J. Radhamani and K. P. S. Chandel. 1991. Preliminary observations on the cryopreservation of desiccated embryonic axes of tea (*Camellia sinensis* L.) O. Kuntze) seeds for genetic conservation. *Cryo Letters* 12:31-36.

Chaudhury, R. and K. P. S. Chandel .1995. Cryopreservation of embryonic axes of Almond seeds. *Cryo Letters*.16 : 51-56.

Cheng, H., G. Zheng and K. L. Tao. 1990. Effects of ultradrying on ageing, cell ultrastructure and vigor of Chinese cabbage seed. *FAO/IBPGR Plant Genet. Resour. Newsl.* 83/84:9-14

Cheng, H., C. Zheng and K. L. Tao. 1998. Effects of ultra drying on ageing, cell ultrasturctures and vigor of Chinese cabbage seed. *FAO/IBPGR Plant Genet. Resour. Newsl.* 83/84:9-14.

Chin, H. F. 1975. Germination and storage of rambutan (*Nephelium lappaceum*) seeds. Malay. *Agric. Res.*, 4: 173-180.

Chin, H. F. 1978. Production and storage of recalcitrant seeds in the tropics. *Acta Hort.* 83:17-22.

Chin, H. F. 1989. Recalcitrant seeds. Food and Fertilizer Technology Centre, *Extension Bulletin No. 288,* Taipei city. pp. 1-17.

Chin, H. F. 1994. Seed Banks: Conserving the past for the future. *Seed Sci. & Technol.*, 22: 385-400.

Chin, H. F. and E. H. Roberts. 1980. Recalcitrant crop seeds. Tropical Press SDN BHD, Kuala Lumpur, Malaysia.

Chin, H. F., M. Aziz, B. B. Ang and S. Hamzah. 1981. The effect of moisture and temperature on the ultrastructure and viability of seeds of *Hevea brasiliensis*. *Seed Sci. & Technol.*, 9: 411-422.

Chin, H. F., Y. L. Hor and M. B Mohd Lassim. 1984. Identification of recalcitrant seeds. *Seed Sci. & Technol.*, 12:429-436.

Chin, H. F., B. Krishnapillay and Y. L. Hor. 1989. A note on the cryopreservation of embryos from young coconuts (*Cocos nucifera* var. Mawa). *Pertanika* 12:183-186.

Cibele, C. M., L. A. B. Marilene and Joao Nakagawa. 2003. Desiccation effects on germination and vigor of King Palm seeds. *Hortic. Bras.* 21(1) : 10.1590/50102-05362003000100019

Clatterbuck, M. K. and F. T. Bonner. 1985. Utilization of food reserves in *Quercus* seed during storage. *Seed Sci. & Technol.*, 13: 121-128.

Clegg, J. S. 1979. Metabolism and the intracellular environment. The vicinal water network model. In: Cell – Associated Water. (Eds: W. Drost-Hansen and J.S. Clegg). Academic Press, New York. PP.363-413.

Clegg, J. S. 1986. The physical properties and metabolic status of Artemia cysts at low water contents: the water replacement hypothesis. In Membranes, Metabolism and Dry Organisms (Ed.) A. C. Leopold. Cornell Univ. Press. Ithaca, N.Y.: 169-187.

Close, T. J., A .A. Kortt and P. M. Chandler. 1989. A cDNA – based comparison of dehydration- induced proteins (dehydrins) in barley and corn. *Plant Mol. Biol.*, 13: 95-108.

Copeland, L. O. and M. B. McDonald. 1995. Principles of Seed Science and Technology. 3rd Edn., Chapman and Hill, New York.

Corbineau, F. and D. Come. 1988. Storage of recalcitrant seeds of four tropical species. *Seed Sci. Technol.* 16: 97-103.

Crowe, J. H. and L. M. Crowe. 1986. Stabilization of membranes in anhydrobiotc organisms. In Membranes, Metabolism and Dry Organisms Ed. A.C.Leopold. Cornell Univ. Press. Ithaca, N.Y.: 188-209.

Das, N. K. and A. K. Ray. 1985. Effect of moisture content changes on arecanut (*Areca catechu*) sprouting behaviour. *Seed Sci. & Technol.*, 13: 861-869.

Daws, M. I., N. C. Garwood and H. W. Pritchard .2006. Prediction of desiccation sensitivity in seeds of woody species ; a probabilistic model based on two seed traits and 104 species. *Annals of Botany* 97 : 667-674.

Dehgan, B. and B. Schutzman. 1989. Embryo development and germination of Cycas seeds. *J. Am. Soc. Hort. Sci.*, 114: 125-129.

Delgado, L. F. and C. J. Barbedo. 2007. Desiccation tolerance of seeds of species of Eugenia. Preq. Argopec. Bras. 42(2).DOI: 1590/90100-204 x 2007000200016.

Deltour, R. 1985. Nuclear activiation auring early germination of the higher plant embryos. *J. of Cell Sci.* 75: 43-83.

Deon Erdey and Craig McGill.2006. ISTA Moisture Committee Progress Report: Working group for assessment of water content in recalcitrant and other non-orthodox seed material. Seed Testing International No.131 April 2006 : 20-21.

Dereuddre, J., S. B landin and N. Hassen.1991. Resistance of alginate coated somatic embryos of carrot to desiccation and freezing in liquid Nitrogen.: 1 Effects of preculture. *Cryo.Letters* 12:125-134.

Dhruv Kumar Misra, Devendra Kumar and R. N. Misra. 1998. Effect of substratum, storage temperature and orientation on percentage seed germination of *Azadirachta indica* A Juss. (Neem). *Seed Res.* 26(1):39-42.

Dickie, J.B., M. J. Balick and I. M. Linington. 1992. Experimental investigations into the feasibility and preservation of palm seeds : An alternative strategy for biological conservation of this economically important plant family. *Biodiversity and Conservation*. 1:112-119.

Dobbs, R. C., D. G. W. Edwards, J. Konishi and D. Wallinger. 1976. Guideline to collecting cones of B.C. Conifers. British Columbia For. Ser. / *Canadian For. Ser. Joint Rep.* No. 3. : 12-18.

Dodd, M. C. and J. Van Staden. 1981. Germination and viability studies on the seeds of *Podocarpus henkelii* Stapf. *S. Afr. J. Sci.*, 77: 171-174.

Doijode, S. D. 1990. Influence of seed moisture on longevity of mango (*Mangifera indica* L.) seeds. *Proc. International Satellite Symposium on Seed Science and Technology*. CCS Haryana Agrl. Univ., : Pp 93-96.

Doijode, S. D. 2001. Seed storage of horticultural crops. (Ed.) S. D. Doijode. The Haworth Press. Inc. Binghampton.N.Y.

Durand, B. J., L. Orcan, V. Yanko, G. Zauberman and Y. Fuchs. 1984. Effects of waxing on moisture loss and ripening of 'Furete' avocado fruit. *Hort. Sci.*, 19: 421-422.

Dure, L. S. III, M. Crouch, J. J. Harada, T. H. D. Ho, J. Mundy, R. Quatrano, T. Tamas and Z.R. Sung. 1989. Common amino acid sequence domains among the LEA proteins of higher plants. *Plant Mol. Biol.*, 12: 475-486.

Dure, L. S. III. 1993. The LEA proteins of Higher Plants. In: Control of Plant Gene Expression (Ed. D. P. S. Verma). CRC Press, Boca Raton, FL, USA, CRC Press : Pp. 325-335.

Dusset, S. F. Engelmann, J. Louarn and M. Noirot. 2004. Inheritance of seed desiccation sensitivity in coffee interspecific cross : evidence of polygenic determination. *J. Exp. Bot.* 55(402):1541-1547.

Eggers Sharon Kim. 2010. Sub imbibed storage of recalcitrant seeds of four species. M.Sc.Thesis. University of KwaZulu-Natal, 2007. http:// hdl. handle. net / 10413/1242.

Ellis, R. H., T. D. Hong and E. H. Roberts. 1985. Handbook of Seed Technology for Genebanks. Vol. 1. Principles and methodology. IBPGR, Rome.

Ellis, R. H., T. D. Hong, E. H. Roberts and K. L. Tao. 1990. Low moisture content limits to relations between seed longevity and moisture. *Ann. of Bot.* 65 : 493-504.

Ellis, R. H. and T. D. Hong. 2006. Temperature sensitivity of the low-moisture-content limit to negative seed longevity-moisture content relations in hermetic storage.*Annals of Botany*, 97, 785-791.

Engelman, F., D. Dumet, N. Chabrillange, A. Abdelnuor-Esquivel, B. Assy-Bah, J. Dereuddre and Y. Duval. 1995. Factors affecting the cryopreservation of coffee, coconut and Oilpalm embryos. *Plant Genetic Resources Newsletter*, 103:27-31.

Engelmann, F., D. Dument, N. Chabrillange, A. Abdelnour-Esquivel, B. Assy-Bah, J. Dereuddre and Y. Duval. 1995. Cryopreservation of zygotic and somatic embryos from recalcitrant and intermediate seed species. IPGRI/ *FAO Plant Gen. Res. Newsl.*, 103: 27-31.

Esau, K. 1977. Anatomy of Seed Plants. Wiley and Sons. New York.

Evans, H. 1953. The preservation of cocoa seeds for transport purposes. In: A Report on Cocoa Research, 1945-1951, Imperial College of Tropical Agriculture, St. Aaugustine, Trinidad. p. 79.

Ezumah. 1986. Germination and storage of Neem seed. *Seed Sci. & Technol.* 14(3).

Faria, J. M. R., J. Buitink, A. A. M. Lammeren Van and H. W. M. Hilhorst.2005. *J. Expt. Bot.* (56) 4180:2119-2130.

Farnsworth, E. 2000. The ecology and physiology of viviparous and recalcitrant seeds. Ann. Rev. Ecol. Syst. 31 : 107-138.

Farrant, J. M. P. Berjak, J. G. M. Cuttin and N. W. Pammenter. 1993. The role of plant growth regulators in the development and germination of the desiccation-sensitive (recalcitrant) seeds of *Avicennia marina. Seed Sci. Res.,* 3 : 55-63.

Farrant, J. M., N. W. Pammenter and P. Berjak. 1988. Recalcitrance – A current assessment. *Seed Sci. & Technol.,* 16: 155-166.

Farrant, J. M., N. W. Pammenter and P. Berjak. 1993. Seed development in relation to desiccation tolerance: a comparison between desiccation – sensitive (recalcitrant) seeds of *Avicenia marina* and desiccation tolerant types. *Seed Sci. Res.,* 3: 1-13.

Farrant, J. M., N. W. Pammenter and P. Berjak.1986. The increasing desiccation sensitivity of recalcitrant *Avicennia marina* seeds with storage time. *Physiologia Plantarum,* 67 : 291-298.

Farrant, J. M., N. W. Pammenter, P. Berjak and C. Walters. 1997. Subcellular organization and metabolic activity during development of seeds that attain different levels of desiccation tolerance. *Seed Sci. Res.* 7 : 135-144.

Farrant, J. M., P. Berjak and N. W. Pammenter. 1992. Proteins in development and germination of a desiccation sensitive (recalcitrant) seed species. *Plant Growth Reg.,* 11: 257-265.

Fernandez, P. G. 1982. Effect of air drying, moist storage and soaking on physiology of jackfruit (*Artocarpus heterophyllus*) seeds. Laguna Coll. *Tech. Bull.* : 45

Finch-Savage, W. E. 1996. The role of developmental studies in research on recalcitrant and intermediate seeds. In: Intermediate / recalcitrant tropical forest tree seeds. (Eds) A. S. Quedraogo, K. Poulsen and F. Stubsgaard. Proc. of workshop on improved methods for handling and storage of intermediate / recalcitrant tropical forest tree seeds, 1995, DANIDA Forest Seed Center : 83-87.

Finch-Savage, W. E., G. A. F. Hendry and N. M. Atherton. 1994. Free radical activity and loss of viability during drying of desiccation – sensitive tree seeds. Proc. Roy. Soc. Edinburgh, 102 : 257-260.

Finch-Savage, W. E., J. H. W. Bergervoet, R. J. Bino. 1998. Nuclear replication activity during seed development, dormancy breakage and germination in three tree species : Norway Maple (*Acer platanoides* L), Sycamore (*Acer pseudoplatanus* L) and Cherry (*Prunus avium* L). *J. of Exptl. Bot.* 55-75.

Fox, J., B. Dixon and D. Monk. 1987.Germination in other plant families.Pp 211-223 In Germination of Australian Native Plant Seed (Ed.) P. Langkamp. Inkata press, Melborne.

Friend, R. J. 1964. Some experiments on the storage of cacao in a viable condition. *Papua New Guinea Agric. J.*, 17: 12-18.

Fu, J. R., B. Z. Zhang, X. P. Wang, Y. Z. Qiao and X. L. Huang. 1990. Physiological studies on desiccation, wet storage and cryopreservation of recalcitrant seeds of three fruit species and their excised embryonic axes. *Seed Sci. & Technol.*, 18: 743-754.

Fu, J. R., J. P. Jin, Y. F. Peng and Q. H. Xia. 1994. Desiccation tolerance in two species with recalcitrant seeds: *Clausena lansium* (Lour.) and *Litchi chinensis* (Sonn.). *Seed Sci. Res.*, 4: 257-261.

Fu. J. R., Q. H. Xiz and L. F. Tang.1993. Effects of desiccation on excised embryonic axes of three recalcitrant seed species and studies on cryopreservation. *Seed Sci. & Technol.*, 21:85-95.

Galau, G. A., D. W. Hughes and L. Dure. 1986. Abscisic acid induction of cloned cotton late embryogenesis-abundant (LEA) mRNAs. *Plant Molecular Biology*, 7 : 155-170.

Gee, O. H., R. J. Probert and S. A. Coolbeer. 1994. 'Dehydrin-like' proteins and desiccation tolerance in seeds. *Seed Science Research*, 4 : 135-141.

Georgi, C. D., V. R. Greenstock and G. L. Teik. 1983. Storage of rubber seed. *Turiaiba*, 12: 136-141.

Girija, T. 1998. Physiological investigation on the recalcitrant behaviour of mango (*Mangifera indica* L.) and Rattan (*Calamus* spp.) seeds. Ph.D. Thesis, Tamil Nadu Agricultural University, Coimbatore.

Goor, A. Y. and C. W. Barney. 1976. Forest tree plaing in arid zones (2nd Edition). Ronald Press. NY.

Gunasekaran, M. 1997. Seed technological studies in clove (*Syzygium aromaticum* L), nutmeg (*Myristica fragrans* Houtt.) and rubber (*Hevea brasiliensis* Muell.-Arg.) M.Sc. Thesis. Dept. of SST, TNAU, Coimbatore

Gunasekaran, M. and V. Krishnasamy. 1998. Retention of vigour and viability of rubber (*Hevea brasiliensis* muell-Arg.) seeds. *Neo Botanica*, 6 (1 & 2): 47-53.

Hagenmaier, R. D. and R. A. Baker. 1993. Reduction in gas exchange of citrus fruit by wax coatings. *J. Agric. Food Chem.*, 41: 283-287.

Halma, F. F. and E. Frolich. 1949. Storing avocado seeds and hastening germination. *Calif. Avocado Soc. Yearbook.* 32:136-138.

Hanson, J. 1984. The storage of seeds of tropical tree fruits. In: Crop Genetic Resources: Conservation and Evaluation. (Ed. J.H.W. Holden and J.T. Williams). Allen and Unwin, London. : pp.53-62.

Hanson, J. 1985. Procedures for handling seeds in Genebanks. International Board for Plant Genetic Resources. Rome.

Harman, G. E., B. L. Nedrow, B. E. Clark and L. R. Mattick. 1982. Association of volatile aldehyde production during germination with poor soybean and pea seed quality. *Crop Sci.*, 22: 712-716.

Harrington, J. F. 1972. Seed storage and longevity. In: Seed Biology. III. Insects and Seed Collection, Storage and Seed Testing (Ed. T.T. Kozlowski). Academic Press, New York. : Pp. 145-245.

Harrington, J. F. 1973. Biochemical basis of seed longevity. *Seed Sci., & Technol.*, 1: 453-461.

Hendry, G. A. F., W. E. Finch-Savage, P. C. Thorpe, N. M. Atherton, S. M. Buckland, K. A. Nilsson and W. E. Seel. 1992. Free radical processes and loss of seed viability during desiccation in the recalcitrant species *Quercus robur* L. *New Phytol.*, 122: 273-279.

Heydecker, W. 1972. Vigour. In: Viability of Seeds. (Ed. E.H. Roberts). Chapman and Hall, London. : Pp. 209-252.

Hill, J. P. and W. Edwards. 2010. Dispersal of desiccation-sensitive seeds is not coincident with high rainfall in a seasonal trophical forest in Australia. *Biotrophica*. 42(3):271-275.

Hoekstra, F. A., A. M. Haigh, F. A. A. Tatteroo and T. van. Roekel. 1994. Changes in soluble sugars in relation to desiccation tolerance in cauliflower seeds. Seed Sci. Res., 4:143-147.

Hong, T. D. and R. H. Ellis. 1990. A comparison of maturation drying, germination and desiccation tolerance between developing seeds of *Acer pesudoplatanus* L and *Acer platanoides* L. *New Phytologist*. 116:596-598.

Hong, T. D. and R. H. Ellis. 1995. Interspecific variation in seed storage behaviour within two genera. *Citrus Seed Sci. and Technol.*, 26:165-181.

Hong, T. D. and R. H. Ellis. 1996. A protocol to determine the seed storage behaviour. Technical Bulletin No. 1. (J. M. M. Engels and J. Toll. Vol. Eds) International Plant Genetic Resources Institute, Rome, Italy.

Hong, T. D. and R. H. Ellis. 2000. Storage. Chapter 3. In. Tropical Tree Seed Manual (Ed.). Vozzo, J.A., USDA Forest Service. Pp: 125-136.

Hor, Y. L. 1984. Storage of cocoa (*Theobroma cacao*) seeds and changes associated with their deterioration. Ph.D. Thesis, Universiti Pertanian Malaysia, Malaysia.

Hong, T. D., and R. H. Ellis. 1995. Interspecific variation in seed storage behaviour within two genera : Coffea and Citrus. *Seed Sci. & Technol.*, 23: 165-181.

Hong, T. D. S. Linington and R. H. Ellis. 1996. Seed storage behaviour : a Compendium. Handbook on Genebanks No. 4. IPBGR, Rome

Hor, Y. L., H. F. Chin and Mohamed Zain Karim. 1984. The effect of seed moisture and storage temperature on the storability of cocoa (*Theobroma cacao*) seeds. *Seed Sci. & Technol.*, 12: 415-420.

Hor, Y. L., P. C. Standwood and H. F. Chin. 1990. Effects of dehydration on freezing characteristics and survival in liquid nitrogen of three recalcitrant seeds. *Pertanika*, 13: 309-314.

Hrazdina, G and R. A. Jensen. 1992. Spatial organization of enzymes in plant metabolic pathways. *Ann. Rev. Plt. Physiol.*, 43 : 241-267.

IBPGR, 1981. Report of first meeting, IBPGR *Ad hoc* Advisory Committee on Seed storage. Kew, UK Pp: 20-25.

Iida, Y., K. Watabe, H. Kamada and H. Harada. 1992. Effects of abscisic acid on the induction of desiccation tolerance in carrot somatic embryos. *J. Plt. Physiol.*, 140 : 356-360.

Ilijin, W. S. 1957. Drought resistance in plants and physiological processes. *Ann. Rev. Plt.Physiol.* 3 : 341-363.

ISTA. 2007. International Rules for Seed Testing. International Seed Testing Association, Basserdorf, Switzerland.

Ivanov, P. V. and J. S. Zlatanova.1989. Quantative changes in the histone content of the cytoplasm and the nucleus of germinating maize embryo cells. *Plt. Physiol. Biochem.* 27(6):925-930.

Jarvis, B. C. 1975. The role of seed parts in the induction of dormancy in hazel nut (*Corylus avellana* L.). *New Physiologist.* 75:491-494.

Jensen, L. A. 1971. Observations on the viability of Borneo camphor (*Dryobalanops aromatica* Gaetn.). *Proc. Internatl. Seed Test. Assoc.*, 36:141-146.

Johnston, J. C. and E. F. Frolich. 1957. Avocado propagation Circ. Calif. *Agric. Exp. Stat.* 463 : 19.

Jones, N. 1976. Techniques for assessing quality of *Triplochiton scleroxylon* (K.Schum) fruits. Nigeria. Federal Department of Forest Research. Proc. Of the Symposium on Variation and Breeding Systems of *Triplochiton scleroxylon* (Ed.K.Schum).April21-28.

Jose Faria, M. R., J. Buitink, van Lammeren, A. A. M and H. W. M. Hilhorst. 2005. Changes in DNA and microtubules during loss and re-establishment of desiccation tolerance in germinating *Medicago truncatula* seeds. *J. of Exptl. Bot.* 56 (418) : 2119-2130.

Kahn, V. 1983. Multiple effects of hydrogen peroxide on the activity of avocado polyphenol oxidase. *Phytochemistry*, 22: 2155-2159.

Kamra, S. K. 1976.Use of x ray radiography for studying seed quality in tropical forestry. Studia Forest Suecica. 131 : Pp 34.

Kay, B. L., W. I. Graves and J. A. Young. 1988. Long term storage of desert shrub seed. Mojave Revegetation Notes, No.23. University of California. Agronomy and Range Science, Davis.

Kemp, R. H. 1975. Seed Collection : Temporary storage and transport, documentation, training, safety and supervision. In Report on FAO/ DANIDA training course on Forest Seed Collection and Handling Vol. II. Rome.

Kermode, A. R. 1990. Regulatory mechanisms involved in the transition from seed development to germination. *Critical Reviews in Plant Sciences* 9 : 155-195.

Kim, Y. H. and J. Janick. 1989. ABA and Polyox- encapsulation or high humidity increases survival of somatic embryos of celery. *Hort. Sci.*, 24: 674-676.

King, M. W. and E. H. Roberts. 1979. The storage of recalcitrant seeds. Achievements and Possible Approaches. IBPGR, Rome.

King, M. W. and E. H. Roberts.1980.The desiccation response of seeds of *Citrus limon* L. *Ann. Bot.*, 45 : 489-492.

Kitto, S. L. and J. Janick. 1985. Production of synthetic seeds by encapsulating asexual embryos of carrot. *J. Amerc. Soc. Hort.* Sci. 110 : 277-282.

Koostra, P. T. and J. F. Harrington. 1969. Biochemical effects of age on membrane lipids of *Cucumis sativa* L. seed. Proc. *Internatl. Seed Test. Assoc.*, 34: 329-340.

Koster, K. L. 1991. Glass formation and desiccation tolerance in seeds. *Plant Physiology*, (96) : 302-304.

Koster, K. L. and C. Leopold. 1988. Sugars and desiccation tolerance in seeds. *Plant Physiology*, 88 : 829-832.

Koster, L. K., N. Reisdorph and J. L. Ramsay. 2003. Changing desiccation tolerance of pea embryo protoplasts during germination. *J. of Exptl. Bot.*, 354 (387) : 1607-1614.

Krishnapillay B. 1989. Towards the development of a protocol for cryopreservation of embryos of recalcitrant seed (*Artocarpus heterophyllus* Lam.) Ph.D. Thesis Univ. of Pertanian, Malaysia.

Krishnasamy, V. 1990. Assessment of critical moisture for seed viability in jack. *South Ind. Hort.*, 384:218-219.

Kundu, M. and J. Kachari. 2000. Desiccation sensitivity and recalcitrant behaviour of seeds of *Aquilaria agallocha* Roxb. *Seed Sci. & Technol.*, 28: 755-760.

Leprince, O., A. Werf Van der, R. Deltour and H. Lambers.1992. Respiratory pathways in germinating maize radicles correlated with desiccation tolerance and soluble sugars. *Physiologia Plantarum*, 85 : 581-588.

Leprince, O., J. Buitink and F. A. Hoekstra. 1999. Axes and cotyledons of recalcitrant seeds of *Castanea sativa* Mill. Exhibit contransting responses of respiration to drying in relation to desiccation sensitivity. *J. of Exptl. Bot.*, 50 (338) :1515-1524.

Leprince, O., N. M. Atherton, R. Deltour and G. A. Hendry. 1994. The involvement of respiration in free radical processes during loss of desiccation tolerance in germinating *Zea mays* L. An electron paramagnetic resonance study. *Plant Physiol.*, 104: 1333-1339.

Leprince, O. 1992. From *Avicennia* to *Zizania* : seed recalcitrance in perspective. *Ann. of Bot.*, 101(2) : 213-228.

Li, C. and W. Q. Sun. 1999. Desiccation sensitivity and activities of free radical scavenging enzymes in recalcitrant *Theobroma cacao* seeds. *Seed Sci. Res.*, 9: 209-217.

Liang, Y. H. and W. Q. Sun .2000. Desiccation tolerance of recalcitrant *Theobroma cacao* embryonic axis: the optimal drying rate and its physiological basis. *J. of Exptl. Bot.* 51:1911-1919.

Loomis, W. D. and J. Battaile. 1966. Plant phenolic compounds and the isolation of plant enzymes. *Phytochemistry*, 5: 423-438.

Madusudanan, K. N. and V .Babu. 1994. Recalcitrancy, viability and germination of nutmeg seeds. *J. of Plantation* Crops 22(1):25-29.

Martins, C. C., J. Nakagawa and M. L. Alves Bovi. 2000. Desiccation tolerance of four seed lots from *Euterpe edulis* Mart. *Seed Sci. & Technol.*, 28: 101-113.

McLanghlin, A., G. A. G. Szabo, G. Eisenmann and S. M. Ciani. 1970. Surface changes and the conductance of phospholipid membranes. *Proc. Natl. Acad. Sci.*, 67: 1268-1278.

Morandini, R. 1962. Forest seed handling, equipment and procedures: I. Seed production, collection and extraction. *Unasylva* 15(4): 15-20.

Moreno Diaz de la Espina, S. 1995. Nulear matrix isolate from plant cells. In Structural and Functional Organization of the Nuclear Matrix (Eds) R. Berezney, K. W. Jeon. *Int. Rev. Cytol.*, 162B : 75-139.

Mundy, J. and N. Chua. 1988. Abscisic acid and water- stress induce the expression of a novel rice gene. *The EMBO Journal*, 7 : 2279-2268.

Mycock, D. J. and P. Berjak. 1990. Fungal contaminants associated with several homoiohydrous (recalcitrant) seed species. *Phytophylactica*, 22: 413-418.

Mycock, D. J., P. Berjak and W. E. Finch-Savage. 2000.Effects of desiccation on the subcellular matrix of the embryonic axes of Quercus robur. In Seed Biology: Advances and Applications (Eds) M. Black, K. J. Bradford, J. Vazquez-Ramos. Wallingfod, Oxon, UK : CABI Publishing: Pp. 197-203.

Nagwekar, D. D., P. M. Haldankar, J. C. Rajput and R. T. Gunjate. 1997. Longevity of areca seed nuts as influenced by storage media and storage period. Indian Cocoa, *Arecanut and Spices* J., 21(3): 63-65.

Nambiar, K. K. N. and M. Koti Reddy. 1979. Fungal infection of stored arecanuts. *J. Plantation Crops* 7(1):50-53.

Nautiyal, A. R. and A. N. Purohit. 1985. Seed viability in sal. III. Membrane disruption in ageing seeds of *Shorea robusta. Seed Sci. & Technol.*, 13: 69-76.

Nautiyal, A. R., A. P. Thapliyal and A. N. Purohit. 1985. Seed viability in sal. IV. Protein changes accompanying loss of viability in *Shorea robusta. Seed Sci. & Technol.*, 13: 83-86.

Nkang, A., D. Omokaro and A. Egbe. 2000. Effects of desiccation on the lipid peroxidation and activities of peroxidase and polyphenol oxidase in seeds of *Telfairia occidentalis. Seed Sci. & Technol.*, 28: 1-9.

Normah, M. N. and H. F. Chin. 1989. Ultra structural changes of rubber seeds during imbibed storage. *Seed Sci. & Technol.*, 9: 411-422.

Normah, M. N. and H. F. Chin. 1991. Changes in germination, respiration rate and leachate conductivity during storage of *Hevea* seeds. *Pertanika*, 14(1): 1-6.

Normah, M. N. and M. Vengadasalam. 1992. Effect of moisture content on cryopreservation of *Coffea* and *Vigna* seeds and embryos. *Cryo-Letters* 13:199-208.

Normah, M. N., H. F. Chin and Y. L. Hor. 1986. Desiccation and cryopreservation of embryonic axes of *Hevea brasiliensis* Muell.-Arg. *Pertanika* 9:299-303.

Oliveira, L. M. Q. and I. F. M. Valio. 1992. Effect of moisture content on germination of seeds of *Hancornia speciosa* Gom. (*Apocynaceae*). *Ann. Bot.*, 69: 1-5.

Ong, S. H., A. G. Noor, A. M. Tan and H. Tan. 1983. New *Hevea* germplasm- its introduction and potential. In: Proc. RRIM Planters Conference. Kuala Lumpur, Malaysia. : Pp. 3-17.

Osborne, D. J. and I. I. Boubriak. 1994. DNA and desiccation tolerance. *Seed Sci. Res.*, 4:175-185.

Palanisamy, V. and H. W. Pritchard. 1999. Desiccation and survival of *Castanea sativa* embryonic axes through cryopreservation. National symposium on Forestry Towards 21st Century, Sep 27-28, Tamil Nadu Agricultural University, Coimbatore. :Pp. 132-133.

Pammenter, N., W. Christina, W. Vertucci and P. Berjak.1991. Homeohydrous (Recalcitrant) seeds: Dehydration, the state of water and viability characteristics in *Landlphia kirkii. Plant Physiol.*, 96:1093-1098.

Pammenter, N. W., P. Berjak, J. M. Farrant, M. T. Smith and G. Ross. 1994. Why do stored hydrated recalcitrant seeds die? *Seed Sci. Res.*, 4: 187-191.

Pammenter, N. W., Valerie Greggains, J. I. Kiokeu, J. Wesley-Smith, P. Berjak and W. E. Finch-Savage. 1998. Effects of differential drying rates on viability retention of recalcitrant seeds of *Ekebergia capensis. Seed Sci. Res.*, 8: 463-471.

Pammenter, N. W. and P. Berjak. 1999. A review of recalcitrant seed physiology in relation to desiccation – tolerance mechanisms. *Seed Sci. Res.*, 9: 13-37.

Pammenter, N. W., P. Berjak, J. Wesley-Smith and C. Vander Willigen. 2002. Experimental aspects of drying and recovery. In: M. Black, H. Pritchard. (Eds.). Desiccation and survival in plants: drying without drying. CAB International, Wallingford. : Pp 93-110.

Panggabean, G. 1979. The effect of storage on germination of jackfruit seeds (*Artocarpus heterophyllus*). *Bull. Penelitian Hortikultura* 7:39-42

Parisof, E. 1988. Study of growth rhythm in young mango plants. Part I. Description, germination and storage of polyembryonic mango seeds. *Fruits* (Paris) 43(2): 97-185.

Patil, R. D., R. T. Gunjate and M. J. Salvi. 1986. Effect of storage conditions on viability of mango seed stones. *J. Maharashtra Agric. Univ.*, 11(3): 362.

Pence, V. C. 1990. Cryostorage of embryo axes of several large-seeded tree species. *Cryobiology*, 27: 212-218.

Pence, V. C. 1991. Cryopreservation of immature embryos of *Theobroma cacao. Plant Cell Reports* 10:144-147.

Ponnusamy, A. S., P. Srimathi and K. Vanangamudi. 2007. Three decades of seed research (1972-2007).Seed Center and Dept. of Seed Science and Technology. TNAU, Coimbatore.

Ponnusamy, A. S. 2003. Seed Technological studies in Neem (*Azadiracthta indica* A. Juss.). Ph.D. Thesis, Dept. of Seed Science and Technology, TNAU, Coimbatore.

Poulsen, K. M. 1992. Sensitivity to desiccation and low temperature (-196°C) of embryo axes from acorns of the pedunculate oak (*Quercus robur* L.). *Cryo Letters*, 13: 75-82.

Poulsen, K. M. and E. N. Eriksen. 1992. Physiological aspects of recalcitrance in embryonic axes of *Quercus robur* L seeds. *Seed Sci. Res.*, 2: 215-221.

Pritchard, H. W. and F. G. Prendergast. 1986. Effects of desiccation and cryopreservation on the *in vitro* viability of embryos of the recalcitrant seed species *Araucaria hunsteinii* K. Schum. *Ann. Exp. Bot.* 37:1388-1397.

Pritchard, H. W. and K. R. Manger. 1998. A calorimetric perspective on desiccation stress during preservation procedures with recalcitrant seeds of *Quercus robur* L. *Cryo Letters*, 19 (Supplement 1): 23-30.

Pritchard, H. W., A. J. Haye, W. J. Wright and K. J. Steadman. 1995. A comparative study of seed viability in Inga species: desiccation tolerance in relation to the physical characteristics and chemical composition of the embryo. *Seed Sci. & Technol.*, 23:85-100.

Pritchard, H. W., C. B. Wood, S. Hodges, H. J. Vautier. 2004. 100 seed test for desiccation tolerance and germination : a case study on 8 tropical palm species. *Seed Sci. & Technol.*, 32: 393-403.

Radha, R. K., S. William Decruse and P. N. Krishnan. 2010. Cryopreservation of excised embryonic axis of *Nothapodytes nimmoniana* (Graham) Mebberly – A vulnerable medicinal tree species of the Western Ghats. *Ind. J. of Biotech.* 9 : 435-437.

Radhamani, J. and K. P. S. Chandel. 1992. Cryopreservation of axes of trifoliate orange (*Poncrius trifoliate* (L) Raf.). *Plant Cell Rep.* 11:204-206.

Raja, K., V. Palanisamy and P. Selvaraj. 2004. Emission of volatile compounds from deteriorating Arecanut (*Areca catechu*, L.) seed. *Seed Res.* 32 (2):207-208.

Raja, K., V. Palanisamy and P. Selvaraju. 2001(a). Desiccation sensitivity of recalcitrant curryleaf (*Murraya koenigii* (Linn.) Sprengel) seeds. *Danida Forest Seed Centre (Denmark) Newsletter*, 9: 27-29.

Raja, K., V. Palanisamy and P. Selvaraju. 2002 (a). Effect of wax coating on storage of recalcitrant arecanut (*Areca catechu* L.) seeds. In: Proc. IUFRO Symposium on "Tree Seeds 2002", September 11-15, University of Athens, Chania, Greece. : Pp 140-147.

Raja, K., V. Palanisamy and P. Selvaraju. 2001(b). Desiccation sensitivity of recalcitrant arecanut (*Areca catechu* L.) seeds. *Danida Forest Seed Centre (Denmark) Newsletter*, 9: 24-26.

Raja, K., V. Palanisamy and P. Selvaraju. 2002. Physiological and biochemical changes during desiccation of recalcitrant arecanut (*Areca catechu* L.) seed. In: Proc. IUFRO Symposium on "Tree Seeds 2002", September 11-15, University of Athens, Chania, Greece. : Pp 133-139.

Raja, K., V. Palanisamy, P. Selvaraju and K. A. Shanmugasundaram. 2001. Desiccation sensitivity of avocado (*Persea americana* Mill.) seeds. *Danida Forest Seed Centre (Denmark) Newsletter*, 8: 22-24.

Raja. K. 2001. Seed handling storage and seedling production in arecanut (*Areca catechu* L). Ph.D. Thesis, Tamil Nadu Agricultural University, Coimbatore

Rajeswari Dayal, B. and K. M. Kaveriappa. 2000. Effect of desiccation and temperature on germination and vigour of the seeds of *Hopea parviflora* Beddome and *Hopea ponga* (Dennst.) Mabb. *Seed Sci. & Technol.*, 28: 497-506.

Ravindra, V., S. Shivashankar, T. Muralieedhara Reddy and S. C. Kotur. 2010. Spongy tissue in Alphonso mango III. Radio tracer evidence for increased mobilization of water from mesocarp to seed. *Curr. Sci.* 99 (5) : 571 -573.

Ray, P. K. and S. B. Sharma. 1987. Growth, maturity, germination and storage of litchi seeds. *Sci. Hort.*, 33: 213-221.

Redenbaugh,K., B. D. Paasch, J. W. Nichol, M. E. Kossler, P. R. Viss and K. A. Walker. 1986. Somatic seeds : Encapsulation of asexual plant embryos. *Biotech.* 4 : 497-801.

Rink, G. and R. D. Williams. 1984. Storage technique affects white oak viability. *Tree Planters Notes*, 35(1): 3-5.

Roberts, E. H and M. W. King.1980. The characteristics of recalcitrant seeds. Pp 1-5 in Recalcitrant Crop Seeds (Eds.) H. F. Chin and E. H. Roberts. Trophical Press. Kualalumpur, Malaysia.

Roberts, E. H. 1972. Storage environment and the control of viability. In: Viability of Seeds. (Ed. E.H. Roberts). Chapman and Hall, London. pp. 14-58.

Roberts, E. H. 1973. Predicting the storage life of seeds. *Seed Sci. & Technol.:* 499-514.

Roberts, E. H., M. W. King and R. H. Ellis. 1984. Recalcitrant seeds, their recognition and storage. In: Crop Genetic Resources – Conservation and Evaluation. (Ed. J. H. W. Holden and J. T. Williams). Allen and Unwin, London. : Pp. 38-52.

Rosenberg, L. A., R. W. Rinne. 1988. Protein synthesis during natural and precocious soybean seed (*Glycine max* [L.] Merr.) maturation. *Plant Phisiology*, 87 : 474-478.

Roy, S. K. and D. S. Khurdiya. 1986. Studies on evaporatively cooled zero energy input cool chambers for the storage of horticultural produce. *Ind. Food Packer* 40(60:26-31.

Rühl, G and M. Dambroth. 1989. Investigations in the causes of sensitivity to cold and water stress of tropical seeds, represented by cacao seeds. III.

Ultrastructure of the parenchyma cells in radicles of desiccating cacao seeds. *Landbauforschung Völkenrode* 39:1-15.

Sabale, S. S., H. R. Nadkarni and R. N. Nawale. 1992. Effect of seed storage on germination and seedling establishment in clove. *Indian Cocoa, Arecanut and Spices J.*, 16 (1): 26-28.

Sangakkara, U. R. 1993. Effect of time of harvest and storage conditions on germination of nutmeg (*Myristica fragrans* Houtt.). *J. Agron. & Crop Sci.*, 170: 97-102.

Sargent, J. A., S. S. Mandi and D. J. Osborne. 1981. The loss of desiccation tolerance during germination : an ultrastructural and biochemical approach. *Protoplasma* 105 : 225-239.

Sasaki, S. 1976. The physiology, storage and germination of timber seeds. In: Seed Technology in the Tropics. (Eds. H.F. Chin, I.C. Enoch and R.M. Raja Harun). Universiti Pertanian Malaysia, Kuala Lumpur, Malaysia. Pp. 11-15.

Selvakumar, P. 1999. Studies on seed storage and nursery management in neem (*Azadirachta indica* A. Juss). M.Sc.Thesis. Dept. of Seed Science and Technology, TNAU, Coimbatore.

Senaratna, T. and B. D. Mc Kersie. 1986. Loss of desiccation tolerance during seed germination. A free radical mechanism of injury. In: Membranes, Metabolism and Dry organisms. (Ed. A.C. Leopold). Cornstock Publishing Associates, Ithaca, New York. : Pp. 85-101.

Senaratna, T., B. D. Mc Kersie and A. Borochov. 1987. Desiccation and free radical mediated changes in plant membranes. *J. Exp. Bot.*, 38: 2005-2014.

Seward, B. R. T. 1980. The production, handling and testing of forest tree seed in Zimbabwe - a review of methods and results. *Zimbabwe Bull. For. Res.* No. 8 . Forest Commission, Salisbury.

Shivarama Reddy, L. 1987. Effect of different stages of maturity and post harvest treatments on the seed viability of *Coffea arabica* L. *J. Coffee Res.*, 17(1):14-25.

Shylla Merlin, J. and V. Palanisamy. 2000. Seed viability and storability of jack fruit (*Artocarpus heterophyllus* L.). *Seed Res.*, 28 (2): 166-170.

Simak, M. 1957. The X ray contrast method for seed testing. *Scots Pine (Pinus silvestris) Meddn. St. Skogsforsk Inst.* 47:1-22.

Simon, E. W. 1974. Phospho lipids and plant membrane permeability. *New Phytol.*, 73: 377-420.

Smith, M. T. and P. Berjak. 1995. Deteriorative changes associated with the loss of viability of stored desiccation - tolerant and sensitive seeds. In: *Seed*

Development and Germination. (Eds. J. Kigel and G. Galili). Marcel Dekker Inc., New York. : Pp 701-746.

Soepadmo, E. 1989. Contribution of reproduction biological studies towards the conservation of and development of Malaysian plant genetic resourses. Pp 1-41 in Genetic Resources of Under Utilized Plants in Malaysia (A.H.Zakri Ed.) Malaysian National Commiettee on Plant Genetic Resourses, Malaysia.

Soepadmo, E. and B. K. Eow. 1976. The reproductive biology of *Durio zibenthinus Murr. Gardens' Bull.*, 29: 25-33.

Song, X., Q. Chen, D. Wang and J. Yang. 1984. A study on the principal storage conditions of *Hopea hainanensis* seeds. *Scientia Silvae Sinicae*, 20: 225-236.

Srimathi, P. 1997.Research focus on seed collection, processing and storage of Amla (*Emblica officianalis* Gaertn.), Jamun (*Syzygium cumnii* Skeels) and ber (*Zizypus mauritiana* Lamk). Ph.D. Thesis. Department of Seed Science and Technology, TNAU, Coimbatore.

Srimathi, P. and G. Sastri. 2004. Seed handling techniques in fruit crops. In. Compendium on Seed Science and Technology (Eds.) V. Krishnasamy, A. S. Ponnusamy, P. Balamurugan, P. Srimathi, N. Natrajan and T. S. Raveendran. Dept. of Seed Science and Technology, TNAU, Coimbatore. Pp:284-314.

Srimathi, P., K. Malarkodi and T. V. Karivaratharaju.2001. Storage media and temperature on storability of recalcitrant Jamun (*Syzygium cuminii*) seeds. *Seed Res.* (29) 2:183-188.

Stanwood, P. C. 1983. Cryopreservation of seed germplasm for genetic conservation. In: Cryopreservation of Plant Cells and Organs (K.K. Kartha, ed.). CRC Press, Florida. : Pp. 200-226.

Sudhakara, K., B. N. Nagaraj, A. V. Santhoshkumar, K. K. Sunilkumar and N. K. Vijaykumar. 2000. Studies on the production and storage potential of synthetic seeds in cocoa (*Theobroma cacao* L.). *Seed Res.* 28(2) :119-125.

Sun, W. Q, T. C. Irving and A. C. Leopold. 1994. The role of sugars, vitrification and membrane phase transition in seed desiccation tolerance. *Physiologia Plantarum* 90:621-628.

Sun, W. Q. 2002. Methods for the study of water relations under desiccation strees. In: M. Black, H. Pritchard. (Eds) Desiccation and survival in plants: drying without drying. CAB International, Wallingford. : Pp 47-91.

Sunilkumar, N. K., K. Sudhakara and N. K. Vijayakumar. 2000. An attempt to improve storage life of *Hopea parviflora* seeds through synthetic seed production. *Seed Res.* 28 (2) : 126-130.

Szezotka, Z. 1975. Changes in the viability of protein synthesis in the embryo axes of Northern Red Oak (*Quercus borealis* Midex) and English oak (*Q. robur* L.) during storage under controlled conditions. *Astroretum Kornicle,* 20: 291-297.

Tamari. C. 1976. Phenology and seed storage trials of dipterocarps. Research Pamphlet No. 69, Forest Research Institute, Kepong, Malaysia.

Tang, H. T. and C. Tamari. 1973. Seed description and storage tests of some dipterocarps. *Malaysian Forester,* 36: 38-53.

Tang, L. E. and J. R. Fu.1993. A reasonable suggestion on moist storage of jackfruit (*Artocarpus heterophyllus*) seeds. Acta Scientiarum Natu. University of Sunyatseni 32:111-115.

Tettroo, F. A. A., C. Bomal, F. A. Hoekstra and C. M. Karssen. 1994. Effect of abscisic acid and slow drying on soluble carbohydrate content in developing embryoids of carrot (*Daucus carota* L.) and alfalfa (*Medicago sativa* L.). *Seed Sci. Res.,* 4 : 203-210.

Tobias, M. N. and N. W. Pammenter. 2006. Dehydration kinetics of embryonic axes from desiccation-sensitive seeds: an assessment of descriptive models.

Tompsett, P. B. 1984. Desiccation studies in relation to the storage of *Araucaria* seed. *Ann. Appl. Biol.,* 105: 581-586.

Tompsett, P. B. 1985. The influence of moisture content and temperature on the viability of *Shorea almon, S. robusta* and *S. roxburghii* seed. *Canadian J. For. Res.,* 15: 1074-1079.

Tompsett, P. B. 1986. The effect of desiccation on the viability of dipterocarp seed. In: Report No. 12 of the Federal Forest Research Institute, Vienna. : Pp. 181-202.

Tompsett, P. B. 1987. Desiccation and storage studies on *Dipterocarpus* seeds. *Ann. Appl. Biol.,* 110: 371-379.

Tompsett, P. B. 1992. A review of the literature on storage of Dipterocarp seeds. *Seed Sci. & Technol.,* 20: 251-267.

Tompsett, P. B. 1983. Handling and storage of Agathis and Araucaria seed. *Sylvilcultura Ano.,* VIII No. 30 : 290-293.

Turnbull, J. W. 1975. Seed extraction and cleaning. In. Report on FAO/ DANIDA Training Course on Forest Seed Collection and Handling. Vol. 2, FAO, Rome.

Umboh, M. I. J. 1987. Storage and germination tests on Shorea javanica seeds. *Biotropica.* 1:58-66.

Vanangamudi, K., G. Mani, R. Umanrani and K. T. Parthiban. 2002. Effect of hyropriming on germination improvement in Neem (*Azadirachta indica* A. Juss). *Seed Research.* 30(1):152-154.

Vanitha, C. 2001. Seed technological studies in cocoa (*Theobroma cacao*) cv. Forastero. M.Sc.Thesis. Dept. of Seed Science and Technology, TNAU, Coimbatore.

Vanitha, C., K. Ramamoorthy, A. Vijayakumar and K. Sivasubramaniam. 2005. Moist sand conditioning to minimize loss of viability in cocoa (*Theoborma cacao* L) seed. *NPR* (4) 6:487-491.

Vasquez, W. 2001. Preliminary results on storage of *Calophyllum brasiliensis*. The Project on Handling and Storage of Recalcitrant and Intermediate *Tropical Forest Tree Seeds Newsletter*, 8: 6-7.

Vertucci, C. W. 1993. Towards a unified hypothesis of seed ageing. In Fourth International Workshop on Seeds : Basic and applied aspects of seed biology. (Eds) D.Come and F. Corbineau, Paris, France:739-746.

Vertucci, C. W., P. Berjak, N. W. Pammenter and J. Crane. 1991. Cryopreservation of embryonic axes of an homeohydrous (recalcitrant) seed in relation to calorimetric properties of tissue water. *Cryo-Letters* 12:339-350.

Vertucci, C. W. and J. M. Farrant. 1995. Acquisition and loss of desiccation tolerance. In Seed Development and Germination. (Eds) J. Kigel and G. Gahlili, Marcel Dekker, Inc. New York. : Pp:237-271.

Vertucci, C. W. and E. E. Roos. 1990. Theoretical basis of protocols for seed storage. *Plt. Physiol.* 94: 1019-1023.

Villa, C. L. 1962. Pruebas para ampliar el periodo de viabilidad d' el cacao. *Agricultura Tecnica en Mexico*, 2: 133-136.

Vlase, I. 1970. Contribution to the development of a method of storing acorns of *Quercus robur* L. for long periods. *Revista Padurilorr*, 85: 616-619.

Von Teichman, I. and A.E. van Wyk.1994. Structural aspects and trends in the evolution of recalcitrant seeds in dicotyledons. *Seed Sci. Res.* 4 : 225-239.

Vozzo, J. A.1988. Seed radiography, Materials evaluation. 46, October : 1450-1455.

Walters, C., J. W. Smith, J. Crane, L. M. Hill, P. Chmialarz, N. W. Pammenter and P. Bejak. 2008. Cryopreservation of recalcitrant (i.e. Desiccation – sensitive) seeds. Plant Cryopreservation: A practical guide. Section II. 465-484. DOI 10.1007 / 978-0-387-72276-4_18.

Wechsberg, G. E., C. M. Bray and J. Probert. 1993. The development and longevity and response to priming of Ranunculus scleratus seeds. In. Fourth International Workshop on Seeds : Basic and applied aspects of seed biology. (Eds) D. Come and F. Corebineau. Paris, France : 845-850.

Wesley-Smith, J., C. W. Vertucci, P. Berjak. N. W. Pammenter and J. Crane. 1992. Cryopreservation of desiccation sensitive axes of Camelia sinensis in relation to dehydration, freezing are and the thermal properties of tissue water. *J. Plant Physiol.,* 140 : 596-604.

Williams, R. J. and A. Leopold. 1989. The glassy state in corn embryos. *Plant Physiology,* 89 : 977-981.

Wills, R. H. H., T. H. Lee, D. Granham, W. B. Mc Glasson and E. G. Hall. 1981. Post harvest. An Introduction to the Physiology and Handling of Fruits and Vegetables. Granada Publishing Ltd., London. : Pp. 115.

Wilson, D. O. Jr. and M. B. Mc Donald. 1986. A convenient volatile aldehyde assay for measuring seed vigor. *Seed Sci. & Technol.,* 14:259-268.

Winters, H. F. 1953. The mangosteen. *Fruit Var. Hort. Digest.,* 8: 57-58.

Woessner, R. A. and K. L. Mc Nabb. 1979. Large scale production of Gmelina arborea Roxb. Seed - a case study. *Commonwealth Forestry Review* 58: 117-121.

Wolfe, J. 1978. Chilling injury in plants. The role of membrane lipid fluidity. *Plant Cell Environ.,* 1: 241-247.

Woodstock, L. W. and R. B. Taylorson. 1981. Soaking injury and its reversal with polyethylene glycol in relation to respiratory metabolism in high and low vigor soybean seeds. *Pl. Physiol.,* 67 : 424.

Yeatman, C. W. and T. C. Neiman. 1978. Safe tree climbing in forest management. *Canadian Forest Service. For. Tech.* Rep. 24.

Young, J. A. and C. G. Young. 1992. Seeds of woody plants in North America. Dioscorides Press. Portland, Oregon.

Zewdie, M. and R. H. Ellis. 1991. Survival of tef and niger seed following exposure to sub-zero temperatures at various moisture contents. *Seed Sci. & Technol.,* 19: 309-317.

Zink, E. and M. Ojima. 1965. The effect of storage conditions on the germinatin capacity of loquat seeds. *Bragantia.* 24:9-12.

□□□

About the Authors

Professor Dr. K. Sivasubramaniam is scientist of many hues. He did is graduate, post graduate and doctorate in the prestigious Tamil Nadu Agriculture University, Coimbatore. With twenty two years spent in seed production, teaching and extension activity he is well known in the field of seed science and has authored five books that have been well received by the peer community. He has also to his credit several schemes and research projects. Presently he is the Professor and Head of Department of Seed Science and Technology, Agricultural College and Research Institute, Madurai, Tamil Nadu.

Dr.K.Raja, Assistant Professor, Seed Science and Technology joined the faculty during 2004. He did his doctorate in recalcitrant seed physiology of arecanut which forms genesis of this book. An avid reader and budding scientist he is an enthusiastic and quite worker. With several international articles to his credit he is currently researching the use of cashew nut shell liquid for controlling bruchids.

Dr. R. Geetha is Associate Professor in the Dept. of Seed Science and Technology at Agricultural College, Madurai. She has a wide range of experience spanning over 15 years. Her knowledge spreads over agricultural and horticultural crops. She has published several scientific articles and manuals for undergraduate courses. Her primary focus is on seed priming techniques.